An Introduction to
ALGEBRAIC STRUCTURES

JOSEPH LANDIN

Professor Emeritus, Mathematics
University of Illinois at Chicago Circle

Dover Publications, Inc.
New York

To Esther

Copyright © 1969 by Joseph Landin.
All rights reserved under Pan American and International Copyright Conventions.

Published in Canada by General Publishing Company, Ltd., 30 Lesmill Road, Don Mills, Toronto, Ontario.
Published in the United Kingdom by Constable and Company, Ltd., 10 Orange Street, London WC2H 7EG.

This Dover edition, first published in 1989, is an unabridged, corrected republication of the work first published by Allyn and Bacon, Inc., Boston, 1969.

Manufactured in the United States of America
Dover Publications, Inc., 31 East 2nd Street, Mineola, N.Y. 11501

Library of Congress Cataloging-in-Publication Data

Landin, Joseph.
    An introduction to algebraic structures / by Joseph Landin.
        p.   cm.
    Reprint. Originally published: New York : Allyn and Bacon, 1969.
    Includes index.
    ISBN 0-486-65940-2
    1. Algebra, Abstract. I. Title. II. Title: Algebraic structures.
QA162.L35   1989
512'.02—dc19   88-7857
                                                          CIP

Preface

THE POWER AND ELEGANCE of the abstract approach in mathematics generally, and in algebra particularly, have been established firmly during this century. The purpose of this book is to acquaint a broad spectrum of students with what is today known as "abstract algebra."

It is widely agreed that the study of mathematics proceeds most effectively from the concrete to the abstract. With this notion as a guide, I have introduced numerous examples with three purposes in view. The first is to base the definitions and theorems upon experience, thus to motivate the development of the abstract theory with elementary arithmetic and geometric considerations. The second is to illustrate the theory with concrete examples in familiar contexts. The third is to give the student intensive computational experience. In keeping with the underlying pedagogical principle, the book begins in a leisurely fashion and the first three chapters are written with sufficient detail to justify the description "easy." Although the fourth and fifth chapters make increasing demands upon the student's mathematical maturity, experience shows that, with a little effort, they remain within his capacity.

The exercises follow a similar pattern. Some are formal and manipulative, tending to illustrate the theory and develop computational skill. Others constitute an integral part of the theory, in that the student is asked to supply proofs or parts of proofs omitted from the text. Still others are intended to stretch mathematical imagination by eliciting both conjectures and proofs.

A few words are in order concerning Chapter 1. As originally conceived, this book was prepared for students who had had a course in sets and numbers adequate to the purposes of Chapter 2 through 5. Chapter 1 was added at a later date to make the book self-contained. The desire for completeness led me to include more in the first chapter than is needed for a study of the subsequent ones. Students who have some knowledge of

iii

elementary set theory and number theory should skip Chapter 1 entirely, referring to it only as occasion requires. Those who have no such knowledge but wish to get on with the subject of algebra as quickly as possible should skip all parts which are not directly pertinent to the rest of the book. I suggest that students in this category omit the following: all proofs in Sections 11 and 13 (many of these will be taken up again in Chapter 2); the discussion of the Completeness Axiom (see footnote 4, page 35); all proofs in Section 14; the entire Section 17, on the complex numbers.

The book was intended as a one-semester (or equivalent) course. In order to complete its main parts within 45-50 hours, some sections, in Chapters 2 through 5, may have to be skipped. I have included several suggestions concerning dispensable parts and no doubt others can be found. The text ends at a point at which Galois Theory could be introduced and covered in another two weeks or so.

The list of persons to whom I am indebted for advice and criticism is unusually long. First of all, there are the participants, some 400, in the Academic Year and Summer Institutes (funded by the National Science Foundation), conducted at the Champaign-Urbana campus of the University of Illinois during the years 1957 to 1964. The early drafts of this text had their trial runs in these Institutes. As the participants were earnest, dedicated and competent high school and college mathematics teachers, their criticisms and reactions were invaluable to me in trying to write a book with broad appeal. For the rest, I should like to mention that Professors Norman Blackburn, Peter Braunfeld, Philip Dwinger, Norman T. Hamilton, and Wilson Zaring, all of the University of Illinois, were most thoughtful and helpful in their comments. I am especially indebted to my wife whose assistance with the chores of preparing this book was indispensable, and whose eye for style and clear, simple expression contributed much to the writing.

JOSEPH LANDIN

Contents

2. The Theory of Groups 52

3. Group Isomorphism and Homomorphism 118

4. The Theory of Rings 143

5. Polynomial Rings 203

1

Sets and Numbers

THE PURPOSE OF this chapter is to introduce two subjects that constitute the foundation of a good deal of higher mathematics, and of algebra in particular. For Chapters 2 through 5 we shall require familiarity with the elements of set theory and the real number system. This initial chapter is devoted to an exposition of the basic concepts and facts of these disciplines. Admittedly our treatment is superficial, but hopefully the reader will find it easy.

I. The Elements of Set Theory

1. *The Concept of Set*

The notions of set theory can be introduced in a rigorous, axiomatic way or, alternatively, in an intuitive fashion. The former method requires an excursion into logic and the foundations of mathematics. The latter enables us to get our show on the road in a quick and relatively painless way. We therefore turn at once to a heuristic description of the concept of *set*.

A *set* is a collection of objects; the nature of the objects is immaterial. The essential characteristic of a set is this: Given an object and a set, then exactly one of the following two statements is true.

(a) The given object is a member of the given set.

(b) The given object is not a member of the given *set*.

EXAMPLES

1. The numbers, 1 and 2, which are solutions of the equation $x^2 - 3x + 2 = 0$ comprise the *solution set* of the given equation. We denote this set by "$\{1, 2\}$."

1

2. The unit circle with center at the origin of the plane is the *set* of points with coordinates (x, y) satisfying the equation $x^2 + y^2 = 1$. For example, the point $(1/2, \sqrt{3/2})$ is in the set, whereas $(1, \sqrt{3/2})$ is not.

3. It might be tempting to speak of "the set of people who will enter the city of Chicago during 2050." But, clearly, such a collection cannot qualify as a set according to our understanding of this term (why?).

DEFINITION 1.   If an object x is a member of a set A, we say x *is an element of* A and write "$x \in A$." If an object y is not an element of a set B, we write "$y \notin B$."

Thus, since 1 is an element of the set $\{1, 2\}$, we write "$1 \in \{1, 2\}$"; since 3 is not an element of $\{1, 2\}$, we write "$3 \notin \{1, 2\}$."

Exercise: Describe a set whose elements are all sets; describe a set whose elements are sets of sets.

In Example 1, Page 1, we denoted the solution set of the equation $x^2 - 3x + 2 = 0$ by "$\{1, 2\}$." This type of notation is convenient in case the elements of the set are few in number. For instance, if the set S consists of the elements a, b, c, d and no others, one writes

$$S = \{a, b, c, d\}.$$

In general, if a set S consists of the elements $a_1, a_2, \ldots, a_n$ where n is a positive integer, then S is denoted by

(1) $$S = \{a_1, a_2, \ldots, a_n\}.$$

While the notation (1) for sets is useful as far as it goes, it will be important to have an additional notation (see Section 5, page 13).

2.   *Constants, Variables and Related Matters*

The words "constant" and "variable" are among the most frequently used terms in mathematics. Since our usages may differ from those the reader is accustomed to, we urge that he read this section carefully.

DEFINITION 2.   A *constant* is a proper name, i.e., a name of a particular thing. Further, a constant *names* or *denotes* the thing of which it is a name.

EXAMPLES
1. "2" is a constant. It is the name of a particular mathematical object—a number.
2. "New York" is a constant. It is a name of one of the fifty states comprising the United States of America.

A given object may have different names, and therefore different constants may denote the same thing.

3. "1 + 1" and "8 · $\frac{1}{4}$" are also constants, both denoting the number two.

4. New York is also known as the "Empire State." Thus "Empire State" and "New York" are names, both denoting the same geographical entity.

Variables occur in daily life as well as in mathematics. We may clarify their use by drawing upon a type of experience shared by almost all people. Various official documents contain expressions such as

(2)   I, _____, do solemnly swear (or affirm) that . . . .

What is the purpose of the "_____" in (2)? Clearly, it is intended to hold a place in which a name, i.e., a constant, may be inserted. The variable in mathematics plays exactly the same role as does the "_____" in (2); it, too, holds a place in which constants may be inserted. However, a "_____" is clumsy for mathematical purposes. Therefore, the mathematician uses an easily written symbol, e.g., a letter of some alphabet, as a place-holder for constants. A mathematician would write (2) as, say,

(3)   I, x, do solemnly swear (or affirm) that . . . ,

and the "x" is interpreted as holding a place in which a name may be inserted.

DEFINITION 3. A *variable* is a symbol that holds a place for constants.

What are the constants that are permitted to replace a variable in a particular discussion? Usually an agreement is made, or understood, as to what constants are admissible as replacements. If an expression such as (2) or (3) occurs in an official document, the laws under which the document is prepared will specify the persons who may execute it. These are the individuals who may replace the variable by their names. Thus, with this variable is associated a *set* of persons, and the names of the persons in the set are the allowable replacements for the variable.

DEFINITION 4. The *range* of a variable is the set of elements whose names are allowable replacements for the given variable.

We have said that letters are to be used as variables. It will also happen that letters will occur as constants; context will make clear whether a constant or a variable is intended.

Variables occur frequently together with certain expressions called "quantifiers." As the term implies, quantifiers deal with "how many." We shall use two quantifiers and illustrate the first as follows:

Let x be a variable whose range is the set of all real numbers. Consider the sentence

(4)   *For each x*, if x is not zero, then its square is positive.

The meaning of (4) is

> *For each* replacement of x by the name of a real number, if the number named is not zero, then its square is positive.

The quantifier used here is the expression "for each." Clearly, the intention is, when "for each" is used, to say something concerning each and every member of the range of the variable. For this reason we call the expression "for each" the *universal quantifier*.

If in place of (4) we write

(5)   For each y, if y is not zero, then its square is positive,

where the range of y is also the set of all real numbers, then the meanings of (4) and (5) are the same. Similarly, y can be replaced by z or some other suitably chosen[1] symbol without alteration of meaning.

The use of the second quantifier is illustrated by the sentence

(6)   *There exists* an x such that x is greater than five and
               smaller than six,

where the range of x is the set of all real numbers. The meaning of (6) is

> There is *at least one* replacement of x by the name of a real number such that the number named is greater than five and smaller than six.

The expression "there exists" (or, "there is") is the *existential quantifier*. If the variable x is replaced throughout (6) by y or some other properly chosen symbol, then the meaning of the new sentence is the same as that of (6).

> DEFINITION 5.   If an occurrence of a variable is accompanied by a quantifier, that occurrence of the variable is *bound*; otherwise it is *free*.

In mathematical discourse, variables frequently occur free. For instance, one finds discussions beginning

> If x is a nonzero real number, then . . .

or

> Let x be a nonzero real number. Then . . .

In such cases, the entire discussion is understood to be preceded by a quantifier. Thus, in algebra texts one sees statements such as

> Let x be a real number. Then
> $$x + 2 = 2 + x.$$

This is interpreted as meaning

> For each real number x, $x + 2 = 2 + x$.

---

[1] A symbol which has been assigned another meaning would not be suitably chosen.

The practice of beginning a discussion with expressions such as "If x is . . ." or "Let x be . . .," i.e., the practice of using the variable as free, will be adopted in many places throughout this text. Just which of the two quantifiers is intended to precede the discussion will be clear from the context.

We conclude this section with a few remarks concerning *equals*. In this text *equals* is used in the following ways:

(i) $1 + 1 = 2$. This statement asserts that $1 + 1$ and 2 are the same object (in this case, same number).

(ii) For each real number x, $(x-2)^2 = x^2 - 4x + 4$. (ii) asserts that for each replacement of x by a real number, $(x-2)^2$ and $x^2 - 4x + 4$ are the same number.

(iii) There exists a real number x such that $x^2 - 4 = 0$. (iii) asserts that there is a replacement for x by a real number such that $x^2 - 4$ and 0 are the same number.

The well-known properties of equals are:

I.   For each x, $x = x$. (Equals is *reflexive*.)
II.   If $x = y$, then $y = x$. (Equals is *symmetric*.)
III.   If $x = y$ and $y = z$, then $x = z$. (Equals is *transitive*.)

3.   *Subsets and Equality of Sets*

DEFINITION 6.   Let A and B be sets. A is a *subset* (or, *part*) of B if and only if all the elements of A are elements of B. We write "$A \subset B$" and also say "A is contained in B." The symbol "$B \supset A$" is defined as "$A \subset B$"; in words, "B contains A."

Stated in the formal language of set theory, using variables, quantifiers, etc., Definition 6 is: A is a *subset* (or, *part*) of B if and only if for each element x, if $x \in A$, then $x \in B$.

DEFINITION 7. A is a *proper* subset of B if and only if $A \subset B$ and $A \neq B$.

Exercises

1. List several subsets of Z. (Z is the set of all integers.)
2. Is the set $\{2, 4, 6, \pi, \frac{1}{2}\}$ a subset of Z? Why? How about $\{2, 4, 6, \frac{1}{2}\}$?

What is the condition that A not be a subset of B? A is not a subset of B, provided the condition asserted in Definition 6 is violated. But that condition is:

For *each* x, if $x \in A$, then $x \in B$.

Consequently, the condition is violated if there is even *one* exception to it. Therefore we deduce

A is not a subset of B if and only if there exists an element $z \in A$ such that $z \notin B$.

If A is not a subset of B, one writes "$A \not\subset B$" or "$B \not\supset A$."

Exercises

Prove:
1. For each set A, $A \subset A$.
2. If $A = B$, then $A \subset B$ and $B \subset A$.
3. If $A \subset B$ and $B \subset C$, then $A \subset C$.
4. If $A \subset B$ and $A \not\subset C$, then $B \not\subset C$.

Although our development of Set Theory is intended to be intuitive, it is convenient to state explicitly one axiom concerning equality of sets. This axiom is simply the converse of Exercise 2, above.

The Axiom of Extensionality. If A and B are sets and if $A \subset B$ and $B \subset A$, then $A = B$.

To illustrate the use of this axiom, we consider an example:

Let A be the set of all equilateral triangles, and let B be the set of all equiangular triangles. The definitions of A and B are different, yet we are confident that $A = B$. Indeed, if $x \in A$, then x is an equilateral triangle. By certain theorems of elementary geometry we know that x is equiangular, and consequently $x \in B$. Therefore, $A \subset B$. By similar arguments one proves that $B \subset A$. The Axiom of Extensionality now asserts (what is truly reasonable) that $A = B$.

Exercise: Give several illustrations of the use of the Axiom of Extensionality.

4. The Algebra of Sets; The Empty Set

The term "algebra" in the present context may strike the reader as unusual. The ordinary use of this word is related to adding and multiplying real numbers. Here we shall develop a formalism for certain operations with *sets*. The justification for the word "algebra" is that this formalism resembles, in certain superficial ways, the elementary operations with the real numbers.

Let A be the set of all positive integers, and let B be the set of all integers less than eleven. Thus $A = \{1, 2, 3, \ldots\}$, and $B = \{10, 9, 8, \ldots, 0, -1, -2, \ldots\}$. The set of elements that A and B *have in common* is $\{1, 2, 3, \ldots, 10\}$ and is called "the intersection of A and B." More generally,

DEFINITION 8. The *intersection* of sets A and B is the set $A \cap B$, of all elements x such that $x \in A$ and $x \in B$.

THEOREM 1. $A \cap B = B \cap A$, i.e., intersection is commutative.

(*Note:* This theorem is analogous to $a \cdot b = b \cdot a$ for the multiplication of real numbers.)

Proof: We prove this theorem by applying the Axiom of Extensionality; we show that $A \cap B \subset B \cap A$, and $B \cap A \subset A \cap B$, whence the desired result follows.

To prove that $A \cap B \subset B \cap A$, we show that each element $x \in A \cap B$ is an element of $B \cap A$. But if $x \in A \cap B$, then (Definition 7) $x \in A$ and $x \in B$. Hence $x \in B$ and $x \in A$, and therefore $x \in B \cap A$. In short, for each $x \in A \cap B$ we have proved $x \in B \cap A$. Consequently, $A \cap B \subset B \cap A$. Similarly, one shows that $B \cap A \subset A \cap B$. Therefore, by the Axiom of Extensionality, $A \cap B = B \cap A$.

Q.E.D.

The reader will note that in the proof it was tacitly assumed that

$$x \in A \text{ and } x \in B \text{ implies } x \in B \text{ and } x \in A.$$

More generally, if p and q are sentences, then a basic principle used in mathematical reasoning is this:

$$p \text{ and } q \text{ is equivalent with } q \text{ and } p;$$

i.e.,

$$p \text{ and } q \text{ implies } q \text{ and } p$$

and

$$q \text{ and } p \text{ implies } p \text{ and } q.$$

This principle is one of several that is used extensively in all mathematical texts.

EXAMPLE: "The rain is falling and the streets are wet" is equivalent with "The streets are wet and the rain is falling."

Venn diagrams provide a convenient device for picturing sets and relationships among them. The idea is to represent sets by simple plane areas.

If $A \subset B$, this situation may be represented diagrammatically in Figure 1 and so on. If $A \not\subset B$ we have pictures such as Figure 2. In each of the last three diagrams one sees that there is an element (i.e., a point) of A which is not an element (point) of B.

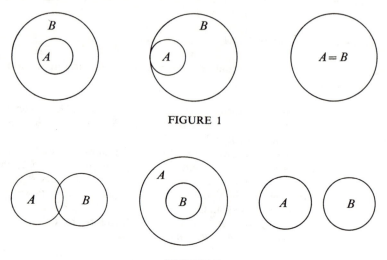

FIGURE 1

FIGURE 2

The intersection, $A \cap B$, of sets A and B is represented as the cross-hatched region in Figure 3. Figure 3 gives added plausibility to the assertion that intersection is commutative.

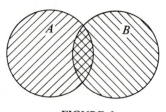

FIGURE 3

At this point, it is natural to ask what is the intersection of sets such as those in Figure 4.

Plainly, these regions have no points in common, and therefore it is reasonable to regard A and B as having no intersection. However, if we take this course, then to some extent the algebra of sets is no longer analogous to the algebra of numbers. In the algebra of numbers, permissible operations with numbers always produce numbers. In a good analogy, one would expect that permissible operations with sets will produce sets. To achieve this objective, we now introduce the concept of the *empty* or *void* set.

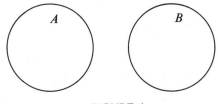

FIGURE 4

DEFINITION 9.   The *empty* (or, *void*) set is the set containing no elements. It is denoted by "$\emptyset$."

We may also define $\emptyset$ as the set such that for each x, $x \notin \emptyset$.
In view of Definition 9, if A and B have no elements in common, then

$$A \cap B = \emptyset,$$

and so an intersection of sets is always a set.
A basic theorem concerning the empty set is

THEOREM 2.   For each set A, $\emptyset \subset A$; i.e., the empty set is a subset of every set.

We shall give an *indirect* proof, that is, a proof by contradiction. In outline, the procedure is:
1. Begin by assuming the theorem is false.
2. If the theorem is false, then its negation is true.
3. Prove that the negation leads to a contradiction of something established previously.
4. Therefore, the negation is false.
5. Consequently, the theorem is true.

Proof of Theorem 2:
    1. Assume that "for each A, $\emptyset \subset A$" is false. Then there is a (at least one) set B, for which "$\emptyset \subset B$" is false.
    2. Hence, there is a set B, for which "$\emptyset \not\subset B$." But if $\emptyset \not\subset B$, there is an element x such that $x \in \emptyset$ and $x \notin B$.
    3. Consequently, there is an element $x \in \emptyset$. This contradicts the definition of $\emptyset$.
    4. Therefore, statement 2 is false.
    5. Hence, the theorem is true.
                                                                Q.E.D.

Exercise: Prove: If $A \subset \emptyset$, then $A = \emptyset$.

THEOREM 3. (Associativity of Intersection). If A, B and C are sets, then

(7) $$(A \cap B) \cap C = A \cap (B \cap C).$$

Exercise: Illustrate equation (7) by means of two Venn diagrams.

A formal proof of Theorem 3 is obtained by means of the Axiom of Extensionality. One proves that $(A \cap B) \cap C \subset A \cap (B \cap C)$, and that $A \cap (B \cap C) \subset (A \cap B) \cap C$, whence the result follows. In turn, the proof depends upon the principle that if p, q and r are sentences, then

$(p$ *and* $q)$ *and* r is equivalent with p *and* $(q$ *and* $r)$.

The details of the proof are left as an exercise.

By repeated applications of Theorem 3 one proves that

$$((A \cap B) \cap C) \cap D = (A \cap B) \cap (C \cap D)$$
$$= A \cap (B \cap (C \cap D)) = \text{etc.,}$$

and so on for any finite number of sets.

DEFINITION 10. $A \cap B \cap C = (A \cap B) \cap C$.

By Theorem 3 we deduce the additional result that $A \cap B \cap C = A \cap (B \cap C)$.

After intersection, the next fundamental operation with sets is *union*. The basic idea is that the union of sets A and B shall consist of all the elements in A and B taken together. More precisely,

DEFINITION 11. The *union* of sets A and B is the set of all x such that $x \in A$ or $x \in B$. We write

$$A \cup B.$$

EXAMPLE: Let $A = \{1, 3, 5, 7, 9\}$, $B = \{2, 4, 6, 8, 10\}$; then $A \cup B = \{1, 2, 3, 4, 5, 6, 7, 8, 9, 10\}$.

Proofs of statements involving union frequently require some rules for the logical term "or." In mathematics and logic the word "or" is used to mean "at least one of the two." Thus in the statement

The sun is shining *or* the streets are wet

at least one, and possibly both, of the statements "the sun is shining," "the streets are wet," is true. In general, if p and q are sentences, then the assertion of the truth of the sentence

$$p \ or \ q$$

means that at least one of p, q (and possibly both) is true. Further, it is a logical principle that

$$p \ or \ q \text{ is equivalent with } q \ or \ p.$$

These few comments are used in proving Exercises 2 and 3 below.

Exercises

1. Verify[2]: If $x \notin A \cup B$, then $x \notin A$ and $x \notin B$.
2. Prove that $A \cup B = B \cup A$, i.e., union is commutative.
3. Prove that $(A \cup B) \cup C = A \cup (B \cup C)$, i.e., union is associative.
4. For each set A, what are $A \cup \emptyset$, $A \cap \emptyset$, $A \cup A$, $A \cap A$?
5. If $A \subset B$, prove that $A \cup B = B$ and $A \cap B = A$. Conversely, show that if $A \cup B = B$, then $A \subset B$; if $A \cap B = A$, then $A \subset B$.
6. For each A and each B, prove that $A \subset A \cup B$ and $A \cap B \subset A$.
7. Using Definition 10 as a model, define $A \cup B \cup C$.
8. Prove that if $B \subset C$, then $A \cup B \subset A \cup C$ and $A \cap B \subset A \cap C$.
9. Prove: If $A \subset C$ and $B \subset C$, then $A \cup B \subset C$. If $A \supset C$ and $B \supset C$, then $A \cap B \supset C$.

In comparing the algebra of sets with the algebra of the real numbers, it is customary to regard *union* as the analog of *sum*, and *intersection* as the analog of *product*. Generally, this comparison works quite well, except that in the algebra of sets there are *two* distributive laws, whereas in the algebra of the real numbers there is but one.

THEOREM 4.   If A, B and C are sets, then
(i) $A \cap (B \cup C) = (A \cap B) \cup (A \cap C)$; intersection is distributive over union, and
(ii) $A \cup (B \cap C) = (A \cup B) \cap (A \cup C)$; union is distributive over intersection.

Proof: (i) We shall apply the Axiom of Extensionality. To prove that $A \cap (B \cup C) \subset (A \cap B) \cup (A \cap C)$, let $x \in A \cap (B \cup C)$. Then $x \in A$, and $x \in B$ or $x \in C$. We distinguish two cases according as (a) $x \in B$, (b) $x \in C$.

CASE (a). $x \in A$ and $x \in B$. Then $x \in A \cap B$. Since (Exercise 6, above) $A \cap B \subset (A \cap B) \cup (A \cap C)$, $x \in (A \cap B) \cup (A \cap C)$.

---

[2] The words "verify" and "show" are used as synonyms for "prove."

CASE (b). $x \in A$ and $x \in C$. Then $x \in A \cap C \subset (A \cap B) \cup (A \cap C)$.

In either case, if $x \in A \cap (B \cup C)$, then $x \in (A \cap B) \cup (A \cap C)$, whence $A \cap (B \cup C) \subset (A \cap B) \cup (A \cap C)$.

On the other hand, since $B \subset B \cup C$, by Exercise 8, $A \cap B \subset A \cap (B \cup C)$. Similarly, $A \cap C \subset A \cap (B \cup C)$. Therefore, by Exercise 9, $(A \cap B) \cup (A \cap C) \subset A \cap (B \cup C)$. Now, using the Axiom of Extensionality, we deduce (i).

(ii)   Exercise.

Exercise: Review the proof of Theorem 4 in the cases that one, two or all three of the sets A, B, C are empty.

A third set-theoretic operation is *complementation*.

DEFINITION 12.   Let A and B be sets. The *complement of B relative to A* is the set of all x such that $x \in A$ and $x \notin B$. It is denoted by "$A - B$."

The Venn diagrams (Figure 5) illustrate three different situations involving relative complements. The shaded areas represent $A - B$.

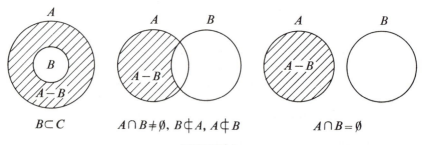

| $B \subset C$ | $A \cap B \neq \emptyset$, $B \not\subset A$, $A \not\subset B$ | $A \cap B = \emptyset$ |

FIGURE 5

Exercises

1. Show that $A - B \subset A$.
2. Prove: $A - A = \emptyset$ and $A - \emptyset = A$.
3. Prove that $A \cap (B - A) = \emptyset$. (Hint: It suffices to show that $A \cap (B - A) \subset \emptyset$.)
4. Prove that $A \cup (B - A) = A \cup B$.
5. Prove: If $A \subset B$, then $A \cup (B - A) = B$.

THEOREM 5.   Let A, B be subsets of E. If $A \cup B = E$ and $A \cap B = \emptyset$, then $B = E - A$.

Proof: Exercise.

COROLLARY. If A is a subset of B, then $B - (B - A) = A$.

Proof: Exercise.

A famous and useful theorem of elementary set theory is

THEOREM 6. (De Morgan). If A, B are subsets of E, then
    (i) $E - (A \cup B) = (E - A) \cap (E - B)$, and
    (ii) $E - (A \cap B) = (E - A) \cup (E - B)$.

The interested reader will find a proof in Hamilton and Landin, *Set Theory and the Structure of Arithmetic*, p. 32, Boston, Allyn and Bacon, 1961.

Exercises

1. Illustrate the meaning of Theorem 6 with Venn diagrams.
2. Prove: $(A - B) \cup (B - A) = (A \cup B) - (A \cap B)$.
3. Prove: $(A - B) \cup (B - C) = (A \cup B) - (C \cap B)$.
4. Let A, B be subsets of E. Prove: $A = \emptyset$ if and only if $B = (A \cap (E - B)) \cup ((E - A) \cap B)$.
5. Let A, B be subsets of E. Show that if $A \subset E - B$, then $A \cap B = \emptyset$ and conversely.
6. If A, B are subsets of E, then $A \supset E - B$ if and only if $A \cup B = E$.

5. *A Notation for Sets*

In Section 1 we introduced a notation which is adequate for sets whose members are not too numerous. Another and more broadly useful notation will be introduced here. The new notation is based upon the idea of a set *consisting of all elements which share a certain property*. While it would be important, in an axiomatic development of set theory, to treat the notion of *property* rigorously, for our purposes it suffices to leave the concept at the intuitive level. We explain the notation by means of examples.

Definition 8 states, "The *intersection* of sets A and B is the set, $A \cap B$, of all elements x such that $x \in A$ and $x \in B$." Here the property in question is

$$x \in A \text{ and } x \in B.$$

We now write the definition thus:

$$A \cap B = \{x \mid x \in A \text{ and } x \in B\}.$$

The braces stand for *the set of*, the stroke, |, for *such that*, and the property is $x \in A$ and $x \in B$.

The set of all countries in Europe is written

$$\{x \mid x \text{ is a country in Europe}\}.$$

Here the property is that of *being a country in Europe*. Consequently

$$\text{Liechtenstein} \in \{x \mid x \text{ is a country in Europe}\},$$

and

$$\text{Peru} \notin \{x \mid x \text{ is a country in Europe}\}.$$

Definitions 10, 11 and 12 are now written in the new notation, thus:

DEFINITION 10'. $A \cap B \cap C = \{x \mid x \in A \cap B \text{ and } x \in C\}.$

DEFINITION 11'. $A \cup B = \{x \mid x \in A \text{ or } x \in B\}.$

DEFINITION 12'. $A - B = \{x \mid x \in A \text{ and } x \notin B\}.$

In general, the new notation is described as

(8) $$\{x \mid P_x\}$$

and read, "the set of all x such that x has the property P_x."

As further examples of the use of the notation (8), we note that the *singleton* set $\{a\}$ is $\{a\} = \{x \mid x = a\}$, and the *doubleton* set $\{a, b\}$ is $\{a, b\} = \{x \mid x = a \text{ or } x = b\}$.

Associated with each set A is another set called the "power set" of A, $P(A)$. We introduce this concept by means of an illustration. Let $A = \{a, b, c\}$ where a, b and c are distinct elements. The complete list of subsets of A is

(9) $\emptyset, \{a\}, \{b\}, \{c\}, \{a, b\}, \{a, c\}, \{b, c\}, \{a, b, c\}.$

The *set of all subsets of A*, $P(A)$, is the set

(10) $\{\emptyset, \{a\}, \{b\}, \{c\}, \{a, b\}, \{a, c\}, \{b, c\}, \{a, b, c\}\}.$

Thus, the elements of the set $P(A)$ defined in (10) are precisely those sets listed in (9).

DEFINITION 13.   The *power set* of a set A is

$$P(A) = \{X \mid X \subset A\}.$$

Exercises

1. Verify the following: (a) $\{x\} \cup \{y\} = \{x, y\}$; (b) $\{x, y\} \cup \{z\} = \{x, y, z\}$; (c) $\{x, y, z\} = \{x, z, y\} = \{z, x, y\} = \text{etc.}$

2. Prove: If $\{x\} = \{y, z\}$, then $x = y = z$.

3. Give an example to show that $\{x, y\} \cap \{y, z\} = \{y\}$ may be false.

4. Suppose that A, B, C, D are sets such that $\{A, B\} = \{C, D\}$. Prove that $A \cap B = C \cap D$ and $A \cup B = C \cup D$. (*Hint:* Consider the two cases, $A = B$ and $A \neq B$.)

5. Let $A = \{a, b, c, d\}$. List all the sets in $P(A)$ in case (a) $c = d$ and a, b, c are distinct and (b) a, b, c, d are all distinct.

6. Generalized Intersection and Union

The definitions of intersection and union given in Section 4 are satisfactory so long as the number of sets involved is rather small. In this section, we give generalized definitions of those operations which are independent of the number of sets involved, and which coincide with our original definitions if we are concerned with only two or three sets. By way of illustration, let $S = \{\{a\}, \{a, b\}, \{a, b, c\}\}$. The intersection of the sets in S is

$$\{a\} \cap \{a, b\} \cap \{a, b, c\} = \{a\}$$

and this set can also be written

(11) $\qquad \{x \,|\, x \in \{a\} \text{ and } x \in \{a, b\} \text{ and } x \in \{a, b, c\}\}$.

It is customary to refer to (11) as "the intersection of all sets T such that $T \in S$" and write

(12) $\qquad \bigcap_{T \in S} T = \{x \,|\, x \in \{a\} \text{ and } x \in \{a, b\} \text{ and } x \in \{a, b, c\}\}$.

Even more simply (12) may be replaced by

(13) $\qquad \bigcap_{T \in S} T = \{x \,|\, x \in T \text{ for each } T \in S\}$.

The virtue of (13) is that it does not require a small number of sets; indeed, one may not even be able to list all the sets whose intersection is desired.

DEFINITION 14. Let S be a set of sets. Then

$$\bigcap_{T \in S} T = \{x \,|\, x \in T \text{ for each } T \in S\}.$$

We also denote "$\bigcap_{T \in S} T$" by "$\bigcap S$" although the former notation is probably more easily understood.

With the same set S as above, consider the union

(14) $\qquad \{a\} \cup \{a, b\} \cup \{a, b, c\} = \{a, b, c\}$,

of all the sets in S; it is called the "union of all sets T such that $T \in S$." As the reader can check, (14) may also be written

$$\{x \,|\, x \in \{a\}\} \cup \{x \,|\, x \in \{a, b\}\} \cup \{x \,|\, x \in \{a, b, c\}\}$$

or

(15) $\{x \mid x \in \{a\}$ or $x \in \{a, b\}$ or $x \in \{a, b, c\}\}$.

Even more simple than (15) is the expression

(16) $\{x \mid$ there exists a $T \in S$ such that $x \in T\}$

and (15) and (16) are clearly the same set.

DEFINITION 15.   Let S be a set of sets. Then

$$\bigcup_{T \in S} T = \{x \mid \text{there exists a } T \in S \text{ such that } x \in T\};$$

$\bigcup_{T \in S} T$ is the *union* of all sets T such that $T \in S$.

The reader will note that Definition 15 does not require a listing of the sets in S, or even a knowledge of how many sets there are. An alternative notation to "$\bigcup_{T \in S} T$" is "$\bigcup S$."

Exercises

1. Let $E = \{a, b, c, d\}$. What is the set $\bigcap_{T \in P(E)} T$? The set $\bigcup_{T \in P(E)} T$?

2. Prove: For all sets E, $\bigcap_{T \in P(E)} T = \emptyset$ and $\bigcup_{T \in P(E)} T = E$.

3. Let A be a set, and let S be a set of sets. Prove: (a) $\bigcap_{B \in S} (A \cup B) = A \cup (\bigcap_{B \in S} B)$;   (b) $\bigcap_{B \in S} (A \cap B) = A \cap (\bigcup_{B \in S} B)$.

4. Let S and T be sets satisfying the following conditions: For each $A \in S$, there is a $B \in T$ such that $A \subset B$. Prove: $\bigcup_{A \in S} A \subset \bigcup_{B \in T} B$.

7.   Ordered Pairs and Cartesian Products

Our initial description of the set concept insisted that the critical feature of a set was this: Given a set A and an element x, then exactly one of the two following statements is true: (i) $x \in A$, (ii) $x \notin A$. Thus a set is simply a collection of elements, no other conditions being required. Along with sets in this sense, we now consider two-element sets in which an *order* is imposed.

DEFINITION 16.   An *ordered pair* (a, b) is a set whose elements are a and b and in which a is the *first component*, b the *second component*. Further,

$$(a, b) = (c, d)$$

if and only if

$$a = c \quad \text{and} \quad b = d.$$

The reader has encountered ordered pairs in his study of analytical geometry. There, it will be recalled, points are located by ordered pairs of real numbers, (x, y). The first component, x, the abscissa, measures the directed distance of the point from the y-axis, and the second component, y, measures its directed distance from the x-axis.

Note that in an ordered pair, (a, b), a and b may be the same element. In analytical geometry, the point with coordinates $(1, 1)$ is the point one unit on the positive side of the y-axis and one unit on the positive side of the x-axis. On the other hand, if $a = b$, then $\{a, b\} = \{a\} = \{b\}$; thus, in this case, "$\{a, b\}$" is another designation for a singleton set.

DEFINITION 17.   The *Cartesian product*, $A \times B$, of sets A and B is the set of all ordered pairs, (x, y), such that $x \in A$ and $y \in B$. Thus,

$$A \times B = \{(x, y) \mid x \in A \text{ and } y \in B\}.$$

The term "Cartesian product" is evidently inspired by analogy with the totality of points of analytical or cartesian geometry.

In Definition 17, it is not required that A and B be different sets; indeed, in the example of analytical geometry, $A = B =$ set of real numbers.

EXAMPLE: If M is a set of men and W is a set of women, then $M \times W$ is the set of all the possible ordered couples that can be formed from these two sets.

Exercise: What is $A \times B$ in case $A = \emptyset$ $(B = \emptyset)$?

From the concept of Cartesian product we derive, in turn, the fundamental notions of *relation, function* (or, *mapping*) and *equivalence relation*. These concepts are among the most important to be studied in this chapter.

Returning to the example cited above, let us agree that an ordered pair (x, y) is a couple if and only if the man, x, is in the relation of *being the husband of* the woman, y. If C is the set of all couples so defined, then $C \subset M \times W$. Thus, from a relation, that of *being the husband of*, one obtains a subset of a Cartesian product. Conversely, from a subset of a Cartesian product, one derives a relation. For instance, suppose A, B, C, D are the children in a given family where A is older than B, B is older than C, and C is older than D. Then, letting $F = \{A, B, C, D\}$,

$$\begin{aligned} F \times F = \{&(A, A), (A, B), (A, C), (A, D), (B, A), \\ &(B, B), (B, C), (B, D), (C, A), (C, B), (C, C), (C, D), \\ &(D, A), (D, B), (D, C), (D, D)\}. \end{aligned}$$

If G is the subset of $F \times F$ defined as

$$G = \{(A, B), (A, C), (A, D), (B, C), (B, D), (C, D)\},$$

then obviously G may be regarded as yielding the relation *is older than*.

Because of the intimate connection between the concept of relation in the ordinary sense and Cartesian product, we make

> **DEFINITION 18.** A *relation C between sets A and B* is a subset of $A \times B$, $C \subset A \times B$. If $B = A$, then C is a *relation on A*. The *domain* of the relation C is the set
>
> $$\mathcal{D}(C) = \{x \,|\, \text{there is a } y \in B \text{ such that } (x, y) \in C\}$$
>
> and the *range* (or, *image*) of C is the set
>
> $$\mathcal{R}(C) = \{y \,|\, \text{there is an } x \in A \text{ such that } (x, y) \in C\}.$$

If C is a relation between A and B, and if $(x, y) \in C$, one writes "$x \, C \, y$"; this may be read, "x is in the relation C to y." If $(x, y) \notin C$, one writes "$x \not\mathrel{C} y$," "x is not in the relation C to y."

Exercises

1. Describe sets A, B and $C \subset A \times B$ so that the relation C corresponds to *is a son of*; what are $\mathcal{D}(C)$ and $\mathcal{R}(C)$?

2. Do the same for the relations *is greater than* and *is less than*.

3. Let $A = B = R$ be the set of all real numbers and let $f \subset A \times B = R \times R$ be the subset defined thus:

$$f = \{(x, y) \,|\, x \in R \text{ and } y = x^2\}.$$

What is the range of the relation f? Give a geometrical interpretation of f.

4. As in Exercise 3, let $A = B = R$, the set of all real numbers, and let $g \subset R \times R$ be defined by $g = \{(x, y) \,|\, x \in R \text{ and } y = 2x\}$. What is the range of g? Give a geometrical interpretation of g.

5. Let M and W be the sets described on page 17. Then $\emptyset \subset M \times W$. Intuitively, what relation does $\emptyset$ describe? Note that, in general, $\emptyset \subset A \times B$, hence $\emptyset$ is a relation. Give an intuitive description of this relation.

Consider once again the example $M =$ a set of men, $W =$ a set of women and $C \subset M \times W$ where $(x, y) \in C$ if and only if x is the husband of y. Clearly, the relation *is the husband of* determines a second relation, namely, *is the wife of*. Consequently, from

$$C = \{(x, y) \,|\, x \in M, \, y \in W \text{ and } x \text{ is the husband of } y\},$$

one obtains the new relation

$$C^{-1} = \{(y, x) \,|\, y \in W, \, x \in M \text{ and } y \text{ is the wife of } x\}.$$

The relation C^{-1} is the *inverse* of the relation C. In general,

> **DEFINITION 19.** Let C be a relation between A and B. The *inverse relation*

of C is the relation $C^{-1} \subset B \times A$ defined by

$$C^{-1} = \{(y, x) \mid (x, y) \in C\}.$$

EXAMPLE: Let $A = \{1, 2, 3, 4\}$, $B = \{2, 4, 6, 8\}$ and $C = \{(1, 2), (2, 4), (3, 6),$
$(4, 8)\}$. Then $C^{-1} = \{(2, 1), (4, 2), (6, 3), (8, 4)\}$.

Exercises

1. Determine the inverse of each of the relations in the set of exercises on page 18.

2. Let $Z = \{\ldots, -3, -2, -1, 0, 1, 2, 3, \ldots\}$ be the set of all integers and define the relation $C \subset Z \times Z$ by

$$C = \{(x, y) \mid (x, y) \in Z \times Z \text{ and } x - y \text{ is divisible by } 3\}.$$

Prove: (a) for each $x \in Z$, $(x, x) \in C$; (b) if $(x, y) \in C$, then $(y, x) \in C$; (c) if (x, y) and (y, z) are in C, then $(x, z) \in C$; (d) $C^{-1} = C$.

8. *Functions (or Mappings)*

In elementary texts, the function concept is usually defined in something like the following way:

> If there is a rule which associates with each value of a variable x in a set of values, one and only one value of a variable y, then y is called a *single-valued function* of x. One writes
>
> $$y = f(x)$$
>
> for each x.

As far as it goes, this informal definition is satisfactory. However, for computational purposes, e.g., the computation of composites of functions or of inverses of functions (where possible), it is advantageous to regard a function as a particular kind of relation. To this end we observe that the essential idea in the above description of function is:

> For each element x in a certain set, there is one and only one element y in some set.

Thus, the function concept has to do with a set of ordered pairs (x, y) such that each x is the first component of exactly one ordered pair in the set. With this discussion as guide, we formulate

DEFINITION 20. Let A and B be nonempty sets. A *function f of A* INTO *B*, or a *mapping* of A INTO B, is a subset f of $A \times B$ such that

(i) for each $x \in A$ there is a $y \in B$ such that $(x, y) \in f$, and

(ii) if $(x, y) \in f$ and $(u, v) \in f$ and if $x = u$, then $y = v$.

Remarks

1. Since $f \subset A \times B$, it follows that a function is a relation. Further, (i) states that the domain of f, $\mathcal{D}(f)$, is the set A and the range $\mathcal{R}(f) \subset B$.

2. The terms "function" and "mapping" are synonymous and will be used interchangeably. A common notation for a mapping f of A into B is "$f: A \longrightarrow B$."

3. Since functions are sets we can deduce equality of functions, say, $f = g$, by proving $f \subset g$ and $g \subset f$ and applying the Axiom of Extensionality.

4. If $f: A \longrightarrow B$ is a mapping, the range of f is also called the "image of f" and is denoted by "$\mathit{Im}\, f$" so that $\mathit{Im}\, f = \mathcal{R}(f)$.

Exercises

1. Which of the relations described in the exercises of pages 18 and 19 are functions? Prove it.

2. Which of the following are relations and which are functions?

(a) $A =$ set of all real numbers, $B =$ set of all nonnegative real numbers, $h = \{(x, x^2) \,|\, x \in A\}$;

(b) $A = B =$ set of nonnegative real numbers and $k = \{(x, x^2) \,|\, x \in A\}$.

(c) $A = B =$ set of real numbers between -1 and $+1$ inclusive, and $F = \{(x, \sqrt{1 - x^2}) \,|\, x \in A\} \cup \{(x, -\sqrt{1 - x^2}) \,|\, x \in A\}$. What is $\mathcal{R}(F)$? What is the graph of F?

(d) With A and B as in part (c), let $G = \{(x, \sqrt{1 - x^2}) \,|\, x \in A\}$. What is the graph of G?

(e) With the same A and B, set $H = \{(x, -\sqrt{1 - x^2}) \,|\, x \in A\}$. What is the graph of H?

(f) Let $A = \{1, 2, 5\}$, $B = \{2, 3, 4, 7\}$ and $K = \{(1, 2), (1, 3), (2, 4), (5, 7)\}$. Is $K - \{(1, 2)\}$ a function?

3. Let $A = B$ be a set, and define $I = \{(x, x) \,|\, x \in A\}$. Prove that I is a mapping. It is defined as *the identity mapping on the set A*. Prove that $I^{-1} = I$.

4. Let $S = \{1, 2, 3\}$, let A be the set $S \times S$, and let $B = \{1, 2, 3, 4, 5, 6\}$. Further, put

$$f = \{((1, 1), 2), ((1, 2), 3), ((1, 3), 4), ((2, 1), 3), ((2, 2), 4),$$
$$((2, 3), 5), ((3, 1), 4), ((3, 2), 5), ((3, 3), 6)\}.$$

Is f a mapping of A into B? Does it remind you of some elementary arithmetic process? Describe f in a simpler way.

5. Let $A = \{1, 2, 3\}$, $C = \{1, 2, 3, 4, 5, 6, 7, 8, 9\}$ and

$$g = \{((1, 1), 1), ((1, 2), 2), ((1, 3), 3), ((2, 1), 2), ((2, 2), 4),$$
$$((3, 1), 3), ((3, 2), 6), ((3, 3), 9)\}.$$

Prove that g is a mapping, $g : A \times A \longrightarrow C$. Describe g as an elementary arithmetic process.

6. Let $Z = \{\ldots, -3, -2, -1, 0, 1, 2, 3, \ldots\}$ be the set of all integers. Verify that the sets

$$s = \{((a, b), c) \,|\, (a, b) \in Z \times Z \text{ and } c = a + b\}$$

and

$$p = \{((a, b), d) \,|\, (a, b) \in Z \times Z \text{ and } d = a \cdot b\}$$

are both mappings of $Z \times Z$ into Z.

7. Let $A = B =$ set of all real numbers and verify that $\alpha = \{(x, x) \,|\, x \in A$ and $x \geq 0\} \cup \{(x, -x) \,|\, x \in A$ and $x < 0\}$ is a function. What is its common name?

It is important to introduce the standard notations and terminology for functions. To this end, consider the example $f : A \to B$ where $A = \{1, 3, 7, 8, 9\}$, $B = \{2, 5, 11, -4\}$ and $f = \{(1, 2), (3, 5), (7, 11), (8, -4), (9, 2)\}$. Then we write $f(1) = 2$, $f(3) = 5$, $f(7) = 11$, $f(8) = -4$, $f(9) = 2$. Thus, "$f(1)$" is simply another designation for 2, "$f(3)$" is another designation for 5, etc. In general, if f is a mapping, $f : A \to B$, and $(x, y) \in f$ where $x \in A$ and $y \in B$, then the second component y of the ordered pair (x, y) is also denoted by "$f(x)$." Therefore we have $y = f(x)$ if and only if $(x, y) \in f$. With this notation, the mapping f may be represented as

$$f = \{(x, f(x)) \,|\, x \in A\}.$$

Again let $f : A \to B$ and let $y = f(x)$ for some $x \in \mathscr{D}(f)$. Then y (i.e., $f(x)$) is the *image* of x. On the other hand, if $y \in \mathscr{R}(f)$ and x is an element in $\mathscr{D}(f)$ such that $(x, y) \in f$, i.e., $y = f(x)$ (there may be more than one), then x is *an inverse image* of y. If C is a subset of $\mathscr{D}(f)$, then the *image* of C is the set

$$f[C] = \{y \,|\, \text{there is an } (x, y) \in f \text{ such that } x \in C\}.$$

Finally, if D is a subset of $\mathscr{R}(f)$, then the *inverse image* of D is the set

$$f^{-1}[D] = \{x \,|\, \text{there is an } (x, y) \in f \text{ such that } y \in D\}.$$

EXAMPLE: Let A be the set of all real numbers, and let B be the set of non-negative real numbers, and set $f = \{(x, x^2) \,|\, x \in A\}$. The image of 0 is 0; the image of 2 is 4; the image of -2 is 4, etc. Both 2 and -2 are inverse images of 4. If $C = \{x \,|\, -1 \leq x \leq 1\}$, then $f[C] = \{y \,|\, 0 \leq y \leq 1\}$. If D is the set of all real numbers greater than or equal to one, then

$$D \subset B \text{ and } f^{-1}[D] = \{x \,|\, x \geq 1\} \cup \{x \,|\, x \leq -1\}.$$

Exercises

In the following exercises, R is the set of all real numbers and Z is the set of all integers.

1. Write the functions $g = \{(x, \ell n\, x) \,|\, x \in R$ and $x > 0\}$ and $h = \{(x, 2x^2 + 3x - 1) \,|\, x \in R\}$ in the customary notations.

2. Define the functions f and g by $f(x) = \dfrac{x^2 - 1}{x - 1}$, $x \in R$ and $x \neq 1$, and $g(x) = x + 1$, $x \in R$, respectively. Is $f = g$?

3. Let $h = \left\{\left(x, \dfrac{x^2 - 1}{x - 1}\right) \middle| x \in R$ and $x \neq 1\right\} \cup \left\{(1, 2)\right\}$ and let g be as in Exercise 2. Is $h = g$?

4. Define $k(x) = \dfrac{x^2 - 1/4}{x - 1/2}$, $x \in Z$, and $m(x) = x + 1/2$, $x \in Z$. Is $m = k$?

5. For the function h in Exercise 1, what are $h^{-1}(-1)$, $h^{-1}(2)$ and $h^{-1}(y)$ for each $y \in \mathscr{R}(h)$?

6. Let f and g be functions such that $\mathscr{D}(f) = \mathscr{D}(g) = A$. Prove that $f = g$ if and only if $f(x) = g(x)$ for each $x \in A$.

7. Suppose f is a function and $x \in \mathscr{D}(f)$. Is there any difference between $f[\{x\}]$ and $f(x)$? If $y \in \mathscr{R}(f)$, what is $f^{-1}[\{y\}]$?

8. Let $f = \{(x, x^2) \,|\, x \in R\}$ and $S = \{x \,|\, -1 \leq x\}$ and $T = \{x \,|\, x \leq 1\}$. What are $f[S \cup T]$, $f[S \cap T]$, $f[S] \cup f[T]$, $f[S] \cap f[T]$?

9. Give an example in which $f^{-1}[D] = \emptyset$.

An important type of mapping already noted in several of the exercises and examples is the *binary operation*. Ordinary addition and multiplication are prototypes of this kind of operation. Thus, Exercises 4, 5 and 6, pages 20 and 21 illustrate the fact that ordinary addition and multiplication can be viewed as mappings. Since other mappings of this type will occur, it is convenient to introduce the concept of binary operation in a general way.

DEFINITION 21.   Let A be a set. A mapping $b : A \times A \longrightarrow A$ is a *binary operation*.

Exercise: Let E be a set. Define union and intersection as binary operations on $P(E)$.

9. *A Classification of Mappings*

Definition 20 admits the possibilities that $\mathscr{R}(f)$ be a proper subset of B or that $\mathscr{R}(f) = B$. If f is a mapping such that $\mathscr{R}(f) = B$, and if we wish to call attention to this fact, we do so by means of some additional terms, namely,

DEFINITION 22.   Let $f : A \longrightarrow B$ be a mapping. If $\mathscr{R}(f) = B$, then f is a mapping of A ONTO B, or, a *surjection*.

Exercise: Among the mappings in Exercises 1-7, pages 20 and 21, pick out all those that are onto, i.e., that are surjections.

Among all mappings, those which are *one-one* occupy an especially important place. Essentially, it is the device used in counting and as such it is probably one of the oldest intellectual methods known to man. At the same time, it is an indispensable tool of sophisticated mathematics.

DEFINITION 23. Let $f: A \longrightarrow B$ be a mapping. f is a *one-one correspondence* if and only if for each $(x_1, y_1) \in f$ and $(x_2, y_2) \in f$, if $y_1 = y_2$, then $x_1 = x_2$.

The term "injection" is also used as a synonym for "one-one correspondence." If f is a one-one correspondence and onto, we say that f is a "one-one correspondence between A and B" or a "bijection."

EXAMPLE: If A, B and f are respectively $\{1, 2, 3, 4, 5\}$, $\{2, 4, 6, 8, 10\}$ and $\{(1, 2), (2, 4), (3, 6), (4, 8), (5, 10)\}$, then f is a one-one correspondence (bijection) between A and B.

Exercises

1. Prove that there is a one-one correspondence between $A \times B$ and $B \times A$. When, if ever, is it true that $A \times B = B \times A$?
2. Prove that there is a bijection $A \times (B \times C) \longrightarrow (A \times B) \times C$.
3. Prove that there is a one-one correspondence between $A \times \{x\}$ and A.
4. Prove: If $\varphi: A \longrightarrow B$ is a bijection, and if $x \notin A$ and $y \notin B$, then $\Psi = \varphi \cup \{(x, y)\}$ is a bijection between $A \cup \{x\}$ and $B \cup \{y\}$ such that $\Psi(x) = y$.
5. Suppose $\varphi: A \longrightarrow B$ is a one-one correspondence, and that $x \in A$. Prove that $\varphi - \{(x, \varphi(x))\}$ is a one-one correspondence between $A - \{x\}$ and $B - \{\varphi(x)\}$.
6. Prove: If φ is a one-one correspondence between A and B, if $x \in A$ and $y \in B$, then there exists a one-one correspondence Ψ between A and B such that $\Psi(x) = y$. (*Hint:* Use Exercises 4 and 5.)
7. Prove that the inverse of a one-one correspondence between A and B is a one-one correspondence between B and A.

10. *Composition of Mappings*

In elementary algebra, along with the basic operations of addition and multiplication, one also studies *substitution* or *composition* of functions. As an illustration, consider the functions

$$f = \{(3, 2), (4, 3), (5, 7)\}, g = \{(2, 1), (3, 9), (7, 14)\} \text{ where } f: A \rightarrow B,$$

$g: C \rightarrow D$ and $A = \{3, 4, 5\}$, $B = \{2, 3, 7\} = C$, and $D = \{1, 9, 14\}$.
Then

$$g(f(3)) = g(2) = 1,$$
$$g(f(4)) = g(3) = 9,$$
$$g(f(5)) = g(7) = 14.$$

Thus, the substitution of the mapping f into the mapping g has yielded the new mapping

$$\{(3, 1), (4, 9), (5, 14)\}.$$

This resultant mapping is the *composite of g and f*. If the given functions are written in tabular form, then the composite, denoted by "$g \circ f$," can be read off directly:

TABLE 1

$g \circ f$	
3	1
4	9
5	14

Definition 24. Let $f: A \rightarrow B$ and $g: C \rightarrow D$ be mappings where $\mathscr{R}(f) \subset C$. The *composite of g and f* is the mapping $g \circ f: A \rightarrow D$ defined by

(17) $g \circ f = \{(x, g(f(x))) \mid x \in A\}.$

By the earlier agreement—that if $(x, y) \in h$ (a mapping), then we write $y = h(x)$—the defining equation (17) becomes

$$g \circ f = \{(x, (g \circ f)(x)) \mid x \in A\},$$

since

(18) $(g \circ f)(x) = g(f(x))$

for each $x \in A$.

According to our definition, $g \circ f$ is not defined unless $\mathscr{R}(f) \subset \mathscr{D}(g)$.

Exercises

1. Let $f = \{(1, 1), (2, 1), (3, 2), (4, 2), (5, 2)\}$ and $g = \{(1, 1), (2, 1), (3, 1)\}$ where $A = B = C = D = \{1, 2, 3, 4, 5\}$. What is $g \circ f$? What is $f \circ g$? Are these composites equal?

2. Let $f: A \rightarrow B$ and $g: C \rightarrow D$ where $\mathscr{R}(f) \subset C$. Prove that $g \circ f$ is a mapping of A into D.

3. If f and g (Exercise 2) are both onto and $B = C$, prove that $g \circ f$ is also onto.

4. Let $f: A \rightarrow B$ be a bijection. Prove that $f^{-1}: B \rightarrow A$ is a bijection.

5. Let $f: A \rightarrow B$ and $g: B \rightarrow C$ be bijections. Prove that $g \circ f$ is a bijection and that $(g \circ f)^{-1} = f^{-1} \circ g^{-1}$.

Exercise 1 above shows that composition of functions is not commutative. However, a most important result is

THEOREM 7.   Composition of mappings is associative, i.e., if $f: A \longrightarrow B$, $g: C \longrightarrow D$, $h: E \longrightarrow F$ where $\mathscr{R}(f) \subset C$ and $\mathscr{R}(g) \subset E$, then

$$h \circ (g \circ f) = (h \circ g) \circ f.$$

Proof: By Exercise 2 above, we know that $h \circ (g \circ f)$ is a mapping of A into F. Similarly, $(h \circ g) \circ f$ is a mapping of A into F.

   To prove that the mappings $h \circ (g \circ f)$ and $(h \circ g) \circ f$ are equal, we need only show (Exercise 6, page 22) that

$$(h \circ (g \circ f))(x) = ((h \circ g) \circ f)(x),$$

for each $x \in A$. Now, for each $x \in A$,

$$(h \circ (g \circ f))(x) = h((g \circ f)(x)) \quad \text{(by Equation (18))}.$$

But $(g \circ f)(x) = g(f(x))$ so that

$$(h \circ (g \circ f))(x) = h(g(f(x))), \; x \in A.$$

Since $f(x) \in C = \mathscr{D}(h \circ g)$, we have

$$h(g(f(x))) = (h \circ g)(f(x)), \; x \in A,$$

and since $h \circ g$ is a mapping of C into F,

$$(h \circ g)(f(x)) = ((h \circ g) \circ f)(x), \; x \in A.$$

This sequence of equations shows that

$$(h \circ (g \circ f))(x) = ((h \circ g) \circ f)(x),$$

for each $x \in A$, and therefore the theorem follows.

Q.E.D.

11.   *Equivalence Relations and Partitions*

In this section we restrict our considerations to relations on a given set. The relations of interest to us are the *equivalence* relations. The reader has already seen examples of such relations without, perhaps, being aware of their special properties.

EXAMPLES
1.   Let T be the set of all triangles, and let $S \subset T \times T$ be the subset consisting of all ordered pairs (x,y) such that x is similar to y. Thus $x \, S \, y$, if and only if x is similar to y. Clearly: (i) for each $x \in T$, $x \, S \, x$, i.e., x is similar to itself; (ii) if $x \, S \, y$, then $y \, S \, x$, i.e., if x is similar to y, then y is similar to x; (iii) if $x \, S \, y$ and $y \, S \, z$, then $x \, S \, z$.

2.   Let R be the set of all real numbers, and let $E \subset R \times R$ such that $(x,y) \in E$ if and only if $x = y$. The properties of equality given on page 5 assert that equality is an equivalence relation.

DEFINITION 25.   An *equivalence relation* on a set A is a relation $\sim$ such that if x, y, z are elements in A, then

(i) $x \sim x$ (i.e., $\sim$ is *reflexive*);

(ii)   if $x \sim y$, then $y \sim x$ ($\sim$ is *symmetric*);

(iii)   if $x \sim y$ and $y \sim z$, then $x \sim z$ ($\sim$ is *transitive*).

Exercises

1. Can an equivalence relation be empty? Give details.

.2. Prove that congruence of triangles is an equivalence relation.

3. Let Z be the set of integers and define: If $x,y \in Z$, then $x \sim_3 y$ if and only if $x - y$ is divisible by 3. Prove that $\sim_3$ is an equivalence relation.

4. With Z as above, and $n \neq 0$, an integer, define: $x \sim_n y$ if and only if $x - y$ is divisible by n. Show that $\sim_n$ is an equivalence relation.

5. With Z as above, define $Z_0 = \{x \mid x \in Z \text{ and } x \sim_3 0\}$, $Z_1 = \{x \mid x \in Z \text{ and } x \sim_3 1\}$, etc. In general, for each integer k let $Z_k = \{x \mid x \in Z \text{ and } x \sim_3 k\}$. Write out descriptions for each of $Z_2, Z_3, Z_4, Z_5, Z_6, Z_{-1}, Z_{-2}, Z_{-3}$. Prove that: (a) each Z_k is nonempty; (b) either $Z_i = Z_j$, or else $Z_i \cap Z_j = \emptyset$; (c) the union of all the Z_k's is the entire set of integers, Z.

6. Show that the inverse of an equivalence relation is the same relation.

7. Let $A = \{a_1, a_2, \ldots, a_{10}\}$, and let U, V, W be the following subsets of A: $U = \{a_1, a_2, a_3\}$, $V = \{a_4, a_5, a_6\}$, $W = \{a_7, a_8, a_9, a_{10}\}$. Define: for each x and y in A, $x \sim y$ if and only if x and y are in the same subset. Verify that $\sim$ is an equivalence relation.

8. Prove that if $A = \{x\}$ is a singleton set, all relations on A are equivalence relations. How many are there?

9. In case A contains more than one element and $\mathscr{D}(\sim) = A$, prove that condition (i) of Definition 25 is a consequence of conditions (ii) and (iii).

Closely related to equivalence relations on a set are the *partitions* of the set. The concept of a partition has been illustrated already in Exercise 5, above, as has its connection with equivalence relation.

DEFINITION 26.   A *partition* P of a set A, $A \neq \emptyset$, is a collection of subsets of A such that

(i) if $S \in P$, then $S \neq \emptyset$;

(ii) if S and T are elements in P, then $S = T$ or $S \cap T = \emptyset$;

(iii) $\bigcup_{S \in P} S = A$.

An element $S \in P$ is an *equivalence class* of the partition P.

The reader should now review Exercises 5, 6 and 7 with great care. The first connection between equivalence relation and partition is that from a partition we can obtain an equivalence relation.

THEOREM 8. Let $P = \{U, V, \ldots\}$ be a partition of A and define $\mathscr{E}(P) \subset A \times A$ thus:

(†) $(x, y) \in \mathscr{E}(P)$ (or, $x \mathscr{E}(P) y$) if and only if there is an $S \in P$ such that $x \in S$ and $y \in S$.

Then $\mathscr{E}(P)$ is an equivalence relation.

Remark: If one writes "$\sim$" in place of "$\mathscr{E}(P)$," then condition (†) becomes

$x \sim y$ if and only if there is an $S \in P$ such that $x \in S$ and $y \in S$.

This is exactly how $\sim$ was defined in Exercise 7, page 26.

Proof of Theorem 8:

If A has fewer than two elements, the proof is left to the reader. Henceforth, assume that A contains at least two elements. By Exercise 9, page 26, it suffices to verify that $\mathscr{E}(P)$ satisfies conditions (ii) and (iii) of Definition 25.

(ii) Suppose $x \mathscr{E}(P) y$. By definition of $\mathscr{E}(P)$, there is an $S \in P$ such that $x \in S$ and $y \in S$. Hence, for the same set S, $y \in S$ and $x \in S$, and therefore (again by definition of $\mathscr{E}(P)$) $y \mathscr{E}(P) x$.

(iii) Let $x \mathscr{E}(P) y$ and $y \mathscr{E}(P) z$. Then there exist sets U and V in P such that x and y are in U, and y and z are in V. Since $y \in U$ and $y \in V$, $U \cap V \neq \emptyset$. But since U and V are in the partition P, condition (ii) of Definition 26 yields that $U = V$. Therefore, x and z are in U $(= V)$ and so $x \mathscr{E}(P) z$.

Consequently, $\mathscr{E}(P)$ is an equivalence relation.

Q.E.D.

We next prove that an equivalence relation yields a partition.

THEOREM 9. Let E be an equivalence relation on the set A. For each $x \in A$, let

$$S_x = \{y \mid y \in A \text{ and } y E x\}.$$

Further, let

$$\mathscr{P}(E) = \{S \mid S \text{ is one of the sets } S_x \text{ for some } x \in A\}.$$

Then $\mathscr{P}(E)$ is a partition of A.

Remark: The technique of proof of this theorem is illustrated in Exercise 5, page 26.

Proof of Theorem 9:

We prove that $\mathscr{P}(E)$ satisfies the three conditions of Definition 26.

(i) Let $S \in \mathscr{P}(E)$. By definition of $\mathscr{P}(E)$, there is an $x \in A$ such that $S = S_x$. However, it is always true that $x \in S_x$ (why?). Therefore, $S \neq \emptyset$ and this verifies the first condition.

(ii) Let S and $T \in \mathscr{P}(E)$. Then there exist x and y in A such that $S = S_x$ and $T = S_y$. We prove that if S and T have even one element in common then $S = T$, and this establishes (ii).

Suppose $z \in S_x \cap S_y$; then $z \in S_x$ and $z \in S_y$ so that by definition of these two sets, $z \, E \, x$ and $z \, E \, y$. Since E is an equivalence relation, $z \, E \, x$ implies $x \, E \, z$. But, $x \, E \, z$ and $z \, E \, y$ yield $x \, E \, y$.

To prove that $(S =)S_x = S_y(= T)$ we show that $S_x \subset S_y$ and $S_y \subset S_x$. Let $u \in S_x$; then $u \, E \, x$ and, since $x \, E \, y$ has been established above, we have $u \, E \, y$. Therefore, $u \in S_y$ and so $S_x \subset S_y$. By similar arguments, one proves that $S_y \subset S_x$. Therefore, $S_x = S_y$, i.e., $S = T$ and (ii) is proved.

(iii) To prove that $\bigcup\limits_{S \,\in\, \mathscr{P}(E)} S = A$, it suffices to show that for each $x \in A$ there is an $S \in \mathscr{P}(E)$ such that $x \in S$. As observed in (i), it is always true that $x \in S_x$, and by definition of $\mathscr{P}(E)$, $S_x \in \mathscr{P}(E)$. This concludes the proof of (iii) and of our theorem.

Q.E.D.

Given a partition P of A, we know that $\mathscr{E}(P)$ is an equivalence relation, and therefore $\mathscr{P}(\mathscr{E}(P))$ is a partition. It can be proved that the latter partition is the one with which we began, namely,

$$\mathscr{P}(\mathscr{E}(P)) = P.$$

Similarly, starting with an equivalence relation E on A, $\mathscr{P}(E)$ is a partition of A and $\mathscr{E}(\mathscr{P}(E))$ is again an equivalence relation on A. Moreover, one can prove that

$$\mathscr{E}(\mathscr{P}(E)) = E.$$

Further, if E_1 and E_2 are equivalence relations on A, then $E_1 = E_2$ if and only if $\mathscr{P}(E_1) = \mathscr{P}(E_2)$. And finally, if P_1 and P_2 are partitions of A, then $P_1 = P_2$ if and only if $\mathscr{E}(P_1) = \mathscr{E}(P_2)$.

Proofs of all the foregoing results will be found in Hamilton and Landin, *Set Theory and the Structure of Arithmetic*, Boston, Allyn and Bacon, 1961. These results justify

DEFINITION 27.   If E is an equivalence relation on a set A, then $\mathscr{P}(E)$ is *the partition determined by E*. If P is a partition of A, then $\mathscr{E}(P)$ is *the equivalence relation determined by P*. An element $S \in \mathscr{P}(E)$ is an *equivalence class determined by the equivalence relation E*.

II.   The Real Numbers

12.   *Introduction*

The common number systems of arithmetic and analysis, as well as certain subsystems and derivable systems, constitute the backbone and a source of inspiration of many algebraic investigations. It therefore behooves us to spend a few hours in the study of these number systems. These systems are:

the real numbers, R;
the rational numbers, Q;
the integers, Z;
the natural numbers, N, and
the complex numbers, C.

Other systems will be studied later, but they will be introduced as the need arises.

The two general methods of studying the foregoing systems are the *constructive* and the *nonconstructive*. In the constructive method one might start with set theory as given and define the natural numbers as certain sets; from the natural numbers, N, construct the integers, Z, and so on. Ultimately one ends up by building the complex numbers, C, from the reals, R. One can easily imagine that this method is rather lengthy and time-consuming. For this reason, and also because certain of the constructive processes are not of immediate interest to us, we shall rely principally, but not exclusively, upon nonconstructive methods.

Our nonconstructive approach will begin with axioms for the real number system, R, and in R we shall designate several subsystems of interest, the natural numbers, the integers and the rational numbers. A construction will be used once in building the complex numbers from the reals.

13. *The Real Numbers*

Individual real numbers, $0, 1, 2, -\sqrt{2}, \pi$, etc., are old acquaintances. What we propose to do is to examine, not the individuals, but rather the real number system as a whole. We shall state general laws of behavior, the *axioms*, from which all the essential properties of R can be deduced. The axioms are given in three sets: I. The Field Axioms; II. The Order Axioms; III. The Completeness Axiom.

The real number system[3] is a set R together with binary operations, $+$, *addition*, and $\cdot$, *multiplication*, satisfying the axioms of I, II and III.

I. *The Field Axioms.*

F.1.　Addition and multiplication are both *commutative*, i.e., if x and y are in R, then $x + y = y + x$ and $xy = yx$.

F.2.　Addition and multiplication are both *associative*; for each x, y and z in R, $x + (y + z) = (x + y) + z$ and $x(yz) = (xy)z$.

---

[3] Properly speaking, the reals consist of R together with the binary operations of $+$ and $\cdot$. For brevity we shall refer simply to "the real numbers, R." Also we adopt, as convenient, the common practice of writing the multiplicative notation as "xy" rather than "$x \cdot y$."

F.3.   There exist unique numbers 0 and 1, $0 \neq 1$, such that for each $x \in R$, $0 + x = x + 0 = x$ and $1 \cdot x = x \cdot 1 = x$.

F.4.   For each $x \in R$, there is a unique $y \in R$ such that $y + x = x + y = 0$.

F.5.   For each nonzero x in R, there is a unique z in R such that $zx = xz = 1$.

F.6.   Multiplication is *distributive* over addition; if x, y, z are elements in R, then $x(y + z) = xy + xz$.

From the Field Axioms we can deduce all those familiar properties of the real numbers which depend only upon addition and multiplication (but not upon *order* and *completeness*). Before stating the axioms of II and III, we shall list a number of these properties as theorems, together with proofs. If time is short the reader should skip these proofs, since they will all be given later in a more general context (see Chapters 2 and 4).

Those who are familiar with the Field Axioms will note that our axioms are redundant. For example, in F.3 it is assumed that there is exactly one *zero* and one *one* having the asserted properties. In a more economical axiom system, it suffices to assume that there are at least one *zero* and one *one* having the given properties. In each case "at most one" can be proved from the axioms. Several other axioms can also be pruned without doing violence to the system as a whole.

DEFINITION 28.   For each $x \in R$, the number y satisfying the condition of F.4 is the *negative* of x and is denoted by "$-x$."

With this definition the equations of F.4 become

$$(-x) + x = x + (-x) = 0.$$

THEOREM 10.   (Cancellation Law for Addition). Let $a, x, y \in R$. Then

$$a + x = a + y$$

if and only if $x = y$.

> *Proof:* If $x = y$, then $a + x = a + y$ is a consequence of the fact that $+$ is a binary operation (Definition 21, page 22).
> Conversely, suppose $a + x = a + y$. Then
>
> $$(-a) + (a+x) = (-a) + (a+y), \qquad \text{(why?)}$$
>
> whence
>
> $$(-a+a) + x = (-a + a) + y; \qquad \text{(by F.2)}$$
>
> therefore
>
> $$0 + x = 0 + y, \qquad \text{(by F.4)}$$
>
> and so
>
> $$x = y. \qquad \text{(by F.3)}$$
>
> Q.E.D.

Using Theorem 10 we can generalize F.4, thus:

THEOREM 11.   For each x and y in R, there is a unique $z \in R$ such that
$$x + z = y.$$

Proof: Given x and y, clearly the number $-x + y$ does the trick; for
$$x + (-x+y) = (x + -x) + y$$
$$= 0 + y$$
$$= y.$$

Hence there is at least one number, namely, $-x + y$, satisfying the given equation.

That there is no more than one such number will now follow from Theorem 10. For, suppose that z and z' are numbers such that
$$x + z = y,$$
and
$$x + z' = y.$$
Then
$$x + z = x + z',$$
whence, by Theorem 10,
$$z = z'.$$
Consequently, $-x + y$ is the *only* number satisfying our equation.
                                                                Q.E.D.

DEFINITION 29.   For each a and b in R, $a - b$ is the number $a + (-b)$.

Observe that the minus sign in "$a - b$" and the minus sign in "$-b$" have different meanings. Note further that $0 - a = -a = 0 + (-a)$.

With the aid of Definition 29, Theorem 11 is now stated: For each x and y in R, there is a unique number z such that
$$x + z = y.$$
The number z is
$$z = y - x.$$

THEOREM 12.   For each $a \in R$, $-(-a) = a$.

Proof: Exercise. (*Hint:* Use F.4 and Definition 28.)

Exercises

1. Use the technique of Theorem 12 to prove that $(-a)b = a(-b) = -(ab)$.
2. Prove that $(-a)(-b) = ab$. Hence deduce $(-1)(-1) = 1$.

3. Prove that $-0 = 0$.

4. Verify the following: (i) $-(a+b) = -a - b$; (ii) $-(a-b) = -a + b$; (iii) $-(-a+b) = a - b$.

DEFINITION 30. For each $a \in R$, $a \neq 0$, the number b such that $b \cdot a = a \cdot b = 1$ is the *reciprocal* of a, denoted by "$1/a$" or by "a^{-1}."

THEOREM 13. (Cancellation Law for Multiplication). Let $a, x, y \in R$ where $a \neq 0$. Then

$$ax = ay$$

if and only if $x = y$.

> *Proof:* If $x = y$, then clearly $ax = ay$. (Why?)
>
> Conversely, if $ax = ay$, then since $a \neq 0$, a has a reciprocal a^{-1} so that
>
> $$a^{-1}(ax) = a^{-1}(ay),$$
>
> hence
>
> $$(a^{-1}a)x = (a^{-1}a)y,$$
>
> or
>
> $$1 \cdot x = 1 \cdot y,$$
>
> whence
>
> $$x = y.$$
>
> Q.E.D.

Using Theorem 13 we generalize F.5, thus:

THEOREM 14. For each x and y in R, $x \neq 0$, there is a unique $z \in R$ such that

$$xz = y.$$

> *Proof:* Given $x \neq 0$ and y, $\dfrac{1}{x}y$ solves the equation. For,
>
> $$x\left(\frac{1}{x} \cdot y\right) = \left(x \cdot \frac{1}{x}\right)y$$
> $$= 1 \cdot y$$
> $$= y.$$

The reader should verify the uniqueness of the solution.

Q.E.D.

THEOREM 15. For each real number a, $0 \cdot a = a \cdot 0 = 0$.

> *Proof:* $a \cdot 1 = a \cdot (1 + 0) = a \cdot 1 + a \cdot 0$. Hence $a \cdot 0$ is a solution of the equation
>
> $$a \cdot 1 + x = a \cdot 1.$$

On the other hand, 0 is also a solution of this same equation. By Theorem 10,

$$a \cdot 0 = 0.$$

Clearly $0 \cdot a = 0$ since $a \cdot 0 = 0 \cdot a$.

Q.E.D.

THEOREM 16.   If a and b are real numbers such that $ab = 0$, then $a = 0$ or $b = 0$.

Proof: If $a = 0$, we're done. (*Note:* In this case, b may also be zero.) If $a \neq 0$, we show that b must be zero. Hence, $a = 0$ or $b = 0$. Now if $a \neq 0$, then a has a reciprocal a^{-1}. Therefore

$$a^{-1}(ab) = (a^{-1}a)b$$
$$= 1 \cdot b$$
$$= b.$$

But

$$a^{-1}(ab) = a^{-1} \cdot 0 = 0,$$

by Theorem 15. Consequently, $b = 0$.

Q.E.D.

DEFINITION 31.   For each a and each b in R, $b \neq 0$, the *quotient* of a by b, a/b, is the number $a \cdot 1/b \,(= (1/b) \cdot a = b^{-1}a = ab^{-1})$. The number a is the *numerator* and b is the *denominator* of the quotient.

In the terminology of Definition 31, the unique solution z of the equation

$$xz = y, \, x \neq 0$$

is the quotient y/x.

We use $\dfrac{\text{``}y\text{''}}{x}$ and "y/x" interchangeably.

Exercises

1. Prove that if $a, b \in R$ and $b \neq 0$, then $-a/b = a/-b = -(a/b)$.
2. Let a, b be nonzero real numbers. Show that $(ab)^{-1} = a^{-1}b^{-1}$.
3. Let $a, b \in R$ and $b \neq 0$. Prove: If x is a nonzero real number, then $ax/bx = a/b$.
4. Let $a,b,c,d \in R$ where $b \neq 0$ and $d \neq 0$. Prove that $a/b \cdot c/d = ac/bd$.
5. Verify that $a/b = c/d$ if and only if $ad = bc$.
6. Assuming $b \neq 0$ and $d \neq 0$, verify the following:

(a) $\dfrac{a}{b} + \dfrac{c}{b} = \dfrac{a + c}{b}$ 　　　　　　　　(b) $\dfrac{a}{b} - \dfrac{c}{b} = \dfrac{a - c}{b}$

(c) $\dfrac{a}{b} + \dfrac{c}{d} = \dfrac{ad + bc}{bd}$ $\qquad\qquad$ (d) $\dfrac{a}{b} - \dfrac{c}{d} = \dfrac{ad - bc}{bd}$

7. The definition of reciprocal of a required that a be nonzero. Could the definition be broadened so as to admit a reciprocal of zero?

II. *The Order Axioms.*

An important property of the real numbers is that they are *ordered* in a way consistent with addition and multiplication. Namely:

The set R of real numbers has a subset P satisfying the axioms:

0.1. For each $x \in R$, exactly one of $x \in P$, $x = 0$, $-x \in P$ holds.

0.2. If x and y are in P, so are $x + y$ and xy.

DEFINITIONS 32. (i) The numbers in P are *positive*. (ii) The numbers in $R - P - \{0\}$ are *negative*. (iii) x *is less than* y if and only if $y - x$ is positive; in this case one writes "$x < y$." (iv) y *is greater than* x, written "$y > x$," if and only if $x < y$. Further we define

$$x \leq y \text{ if and only if } x < y \text{ or } x = y,$$
$$y \geq x \text{ if and only if } y > x \text{ or } y = x.$$

It follows immediately from our definitions that $x > 0$ if and only if x is positive, $x < 0$ if and only if x is negative.

The familiar results below are simple consequences of the axioms and, with few exceptions, the proofs are left to the reader.

THEOREM 17. If a and b are real numbers, then exactly one of $a < b$, $a = b$, $a > b$ holds.

THEOREM 18. If $a < b$ and $b < c$, then $a < c$.

THEOREM 19. If $a < b$, then $a + c < b + c$.

THEOREM 20. If $a < b$ and $0 < c$, then $ac < bc$.

THEOREM 21. If $a \neq 0$, then a^2 is positive.

> *Proof:* Since $a \neq 0$, either $a > 0$ or $a < 0$. In case $a > 0$, $a^2 = a \cdot a > 0$, by 0.2. On the other hand, if $a < 0$, then $-a > 0$ (why?) and therefore $(-a)(-a) > 0$. But since $(-a)(-a) = a \cdot a = a^2$, our theorem follows.
>
> Q.E.D.

THEOREM 22. $0 < 1$.

THEOREM 23. If $a < b$ and $c < 0$, then $ac > bc$.

THEOREM 24.   If $a < b$, then $-a > -b$.

THEOREM 25.   If $a < 0$, then $-a > 0$.

Exercises

Prove the following theorems:
1. The sum of two negative numbers is negative.
2. $a > 0$ if and only if $1/a > 0$; $a < 0$ if and only if $1/a < 0$.
3. $0 < a < b$ if and only if $0 < b^{-1} < a^{-1}$.
4. If $a \leq b$ and $b \leq c$, then $a \leq c$.
5. If $a \leq b$ and $b < c$, then $a < c$; if $a < b$ and $b \leq c$, then $a < c$.
6. There is no real number a such that $a^2 + 1 = 0$.

III. *The Completeness Axiom.*

The eight axioms studied thus far yield all the properties of the real numbers needed for elementary algebra. One further axiom is required, however, to round out our investigation of R. Although this *Completeness Axiom* is of fundamental importance in mathematics generally, in the present introductory study of abstract algebra it plays only an incidental role.[4] Before stating the final axiom, we require some additional concepts.

DEFINITION 33.   Let S be a subset of R. A number b is an *upper bound* for S if and only if, for each $x \in S$, $x \leq b$. A set having an upper bound is *bounded above*.

Some subsets of R possess upper bounds, others do not. However, if a set has an upper bound b, then $b + a$ is also an upper bound for each positive number a.

EXAMPLES:
1. The set of positive integers $\{1, 2, 3, \ldots\}$ clearly does not have an upper bound.
2. The set $\{x \mid x \in R$ and $0 < x < 1\}$ does have an upper bound. Evidently, *one* is an upper bound and so are all real numbers greater than 1.
3. The set of negative numbers has 0 as an upper bound.

Among the upper bounds of the set in Example 2, *one* is obviously a *best* or *closest fitting* or *least upper* bound. It is best in the sense that no number smaller than *one* is an upper bound for the set.

---

[4] In fact, all that is needed of this section is the knowledge that the real numbers, R, are complete, whereas the rational numbers, Q (see Section 16), are not. Consequently, Q is a proper subset of R.

DEFINITION 34.   Let S be a subset of R. A number b is a *least upper bound* or *supremum* for S if and only if

  (i) b is an upper bound for S, and

  (ii) no number smaller than b is an upper bound for S.

One writes "l.u.b." for "least upper bound" and "sup" for "supremum."

Exercises

1. Prove that the empty subset of R is bounded above. Does $\emptyset$ have a l.u.b.? Why?

2. Prove that a set has at most one l.u.b.

Axiom C (Completeness). Every nonempty subset of R which possesses an upper bound has a least upper bound in R.

A major consequence of Axiom C is

THEOREM 26.   (i) If $a > 0$ is a real number and $n > 0$ is an integer, then there is one and only one real number $b > 0$ such that $b^n = a$.

  (ii) If a is negative and $n > 0$ is odd, there is one and only one real number b such that $b^n = a$.

The interested reader will find a proof of Theorem 26 in Hamilton and Landin, *Set Theory and The Structure of Arithmetic*, pages 239–242, Boston, Allyn and Bacon, 1961.

An immediate consequence of the above theorem is the

COROLLARY.   There exists a unique positive real number b such that $b^2 = 2$.

Note: If the positive number b is a square root of two, so is $-b$. One writes "$b = \sqrt{2}$."

As we shall see later (Section 16), the rational numbers Q do *not* contain a square root of two. Since the rationals are defined as a subset of the reals, it follows that Q *is a proper subset of R.*

Exercises

1. Define *lower bound, bounded below, greatest lower bound* (or, *infimum*) by analogy with *upper bound*, etc.

2. Prove that every subset of R which is bounded below has a greatest lower bound in R.

We complete this section with the definition of the *absolute value function* and a set of exercises.

DEFINITION 35. The *absolute value function* is the mapping $\alpha: R \longrightarrow R$ such that

$$\alpha(x) = \begin{cases} x, & \text{if } x \in R \text{ and } x \geq 0, \\ -x, & \text{if } x \in R \text{ and } x < 0. \end{cases}$$

For each $x \in R$, $\alpha(x)$ is the *absolute value* of x.

The standard notation for the absolute value is a pair of vertical bars; one writes

$$|x| = \alpha(x).$$

Henceforth we shall use the standard notation.

Exercises

Prove:
1. For each $x \in R$, $|x| = |-x|$; $x \leq |x|$ and $-|x| \leq x$.
2. If $|x| = |y|$ where $x, y \in R$, then $x = \pm y$.
3. For each x and each y in R, $|xy| = |x| \cdot |y|$.
4. For each x and each a in R where $a \geq 0$, $|x| \leq a$ if and only if $-a \leq x \leq a$.
5. For each x and each y in R, $|x| \leq |y|$ if and only if $-|y| \leq |x| \leq |y|$.
6. For each x and each y in R, $|x + y| \leq |x| + |y|$. (The Triangle Inequality.)
7. For each x and each y in R, $|x - y| \leq |x| + |y|$.
8. For each x and each y in R, $||x| - |y|| \leq |x - y|$.

14. The Natural Numbers

Of all the subsystems of R to be studied here the simplest is N, the system of *natural numbers* (or, *nonnegative integers*). As we know from elementary arithmetic, it consists of the set $\{0, 1, 2, 3, \ldots\}$ together with the operations of addition and multiplication. In order to define the system N formally, we begin by introducing a concept which is characteristic of certain subsets of R.

DEFINITION 36. A subset H of R is *hereditary* if and only if $x \in H$ implies $x + 1 \in H$.

In practice, when we wish to prove that a given set A is hereditary, the first step is (i) to assume that $x \in A$. This step is called the *induction hypothesis*. The second step is (ii) to prove, using (i) and previously established information, that $x + 1 \in A$.

EXAMPLES:
1. The set of all real numbers is hereditary.
2. The empty set is hereditary.
3. From our prior knowledge we know that N itself is hereditary.

THEOREM 27.   An intersection of hereditary sets is hereditary.

Proof: Exercise.

Now let $\mathscr{H}$ be the collection of subsets of R defined thus: $H \in \mathscr{H}$ if and only if
   (i) $0 \in H$, and
   (ii) H is hereditary.

Example 1, above, shows that $\mathscr{H}$ is nonempty. The principal definition of this section is

DEFINITION 37.   The set N of *natural numbers* (or, *nonnegative integers*) is

$$N = \bigcap_{H \in \mathscr{H}} H.$$

Since $0 \in H$ for each $H \in \mathscr{H}$, it follows that $0 \in N$; since N is hereditary (by Theorem 27), $0 + 1 = 1 \in N$; and further, $1 + 1 = 2 \in N$; and $2 + 1 = 3 \in N$, etc. Thus the numbers

$$0, 1, 2, 3, \ldots$$

are certainly elements of N. That N contains no numbers other than these will be a consequence of our investigation.

Note that Axioms F.1, F.2, F.6, 0.1 and 0.2 hold for the numbers in N since they hold for *all* real numbers and since $N \subset R$. Further, since 0 and 1 are in N, it follows that F.3 is also satisfied by N. However, F.4 and F.5 may *not* hold for N since the numbers which exist by these axioms may *not* be elements in N. In fact, we shall prove that F.4 and F.5 are violated by N.

It is important to know that sums and products of natural numbers are again natural numbers. These results can be proved as a consequence of that familiar, basic result known as the "Principle of Finite Induction."

THEOREM 28.   (Principle of Finite Induction, PFI). If M is a hereditary subset of N containing zero, then $M = N$.

Proof: If M is a hereditary subset of N, it is also a hereditary subset of R. Further, since $O \in M$, $M \in \mathscr{H}$ and therefore $N \subset M$. But, by hypothesis, $M \subset N$. Therefore, $M = N$.

Q.E.D.

While our statement of the PFI may not look like the induction principle the reader learned in algebra, it is actually the same principle. It has been stated here in a form making its application in proofs fairly simple. The scheme for using the PFI is as follows:

One wishes to prove that all the natural numbers have a certain property. To this end, one defines

$$M = \{x \mid x \in N \text{ and } x \text{ has the given property}\}.$$

Thus M consists of all elements x such that
(i) x is a natural number, and
(ii) x has the given property.

By (i), $M \subset N$. Now if we can prove that $M = N$, then for each $y \in N$ it must be true that $y \in M$. Hence by (ii), each $y \in N$ must have the given property. Since we already know that $M \subset N$, it suffices to prove that $0 \in M$ and that M is hereditary. It then follows, by the PFI, that $M = N$.

We illustrate the foregoing with

THEOREM 29. A sum of natural numbers is a natural number. In other words, if $n, x \in N$, then $n + x \in N$.

Proof: Let $n \in N$ and set

$$M = \{x \mid x \in N \text{ and } n + x \in N\}.$$

By the above discussion we wish to establish that $0 \in M$ and that M is hereditary.

Since zero is a natural number and $n + 0 = n \in N$, it is immediate that $0 \in M$. Now if $z \in N$, then $z + 1 \in N$ (since N is hereditary). In particular, $n + 1 \in N$, and therefore $1 \in M$.

To prove that M is hereditary, suppose $x \in M$. Then x is a natural number, and $n + x \in N$. We show that $x + 1$ is a natural number, and that $n + (x+1) \in N$. The first of these assertions follows directly from the fact that N is hereditary. As for the second,

$$n + (x + 1) = (n + x) + 1, \qquad \text{(by F.2)}$$

and $n + x \in N$ by the induction hypothesis. Consequently $(n + x) + 1 \in N$, whence $n + (x + 1) \in N$. Therefore, $n + x \in N$ for each $n \in N$ and each $x \in N$.

Q.E.D.

THEOREM 30. A product of natural numbers is a natural number, i.e., if $n, x \in N$, then $nx \in N$.

Proof: Exercise.

We next use the PFI to prove that N contains no negative numbers.

THEOREM 31.   For each $x \in N$, $x = 0$ or $x \geq 1$.

Proof: Let

$$M = \{x \mid x \in N \text{ and } (x = 0 \text{ or } x \geq 1)\};$$

thus M consists of all elements x such that (i) $x \in N$, and (ii) $x = 0$ or $x \geq 1$.

Clearly, $0 \in M$. To prove M hereditary, suppose $n \in M$. Then either $n = 0$, whence $n + 1 = 1$ and so $n + 1 \in M$; or else $n \geq 1$. In the latter case $n + 1 \geq 1 + 1 = 2 > 1$, so $n + 1 \geq 1$, and again $n + 1 \in M$. In any case, M is hereditary; therefore, $M = N$. Consequently, each natural number is either zero or ≥ 1.

Q.E.D.

COROLLARY 1. N contains no negative numbers.

COROLLARY 2.   There is no natural number x such that $0 < x < 1$.

COROLLARY 3.   If $n \in N$, there is no natural number x such that $n < x < n + 1$.

Exercise: Prove the three corollaries.

One of the most important applications of the PFI is in establishing the *Well-Ordering Principle* (WOP) for N.

THEOREM 32.   (Well-Ordering Principle). Every nonempty subset of N contains one and only one smallest element. In other words, if $X \subset N$ and $X \neq \emptyset$, then there is a unique $n \in X$ such that for each $m \in X$, $n \leq m$.

Proof: The reader should verify the following remark: If $X \subset Y \subset N$, and if b is a smallest element in Y, then $b \leq x$ for each $x \in X$.
Let

$$M = \{m \mid m \in N, \text{ and if } A \subset N \text{ and } m \in A,$$
$$\text{then } A \text{ contains a smallest number}\}.$$

First, $0 \in M$. For, if $0 \in A \subset N$, then $0 \leq x$ for each $x \in N$. Consequently, $0 \leq y$ for each $y \in A$. Since $0 \in A$, 0 is a smallest number in A.

Next let $x \in M$. To prove that $x + 1 \in M$, let $x + 1 \in B \subset N$. If $x + 1 \leq y$ for each $y \in B$, we're finished. If not, set

$$C = B \cup \{x\}.$$

Since $x \in C \subset N$, it follows (by the assumption that $x \in M$) that C contains a smallest number, s. Now if $s \in B$, then s is a smallest number in B. If $s \notin B$, then $s = x$ and therefore $x =$

$s < y$ for each $y \in B$. But $x + 1 \in B$ and since there is no natural number z such that $x < z < x + 1$, we deduce that $x + 1 \leq y$ for each $y \in B$. Therefore $x + 1$ is a smallest element in B, hence $x + 1 \in M$. This proves that every nonempty subset of N has a smallest element.

The uniqueness of the smallest element follows thus: If s_1, s_2 are smallest elements of $A \subset N$, then $s_1 \leq s_2$ and $s_2 \leq s_1$. Therefore, $s_1 = s_2$.

Q.E.D.

An important theorem that one usually meets first in elementary geometry is the *Archimedean Law*. In geometric guise the proposition is something like the following: If a and b are lengths, then by repeated application of the length a one can exceed the length b. In the theory of real numbers this result is stated precisely, thus:

THEOREM 33. (The Archimedean Law). If a, b are positive, real numbers, then there is a natural number $n > 0$ such that $na > b$.

Proof: By contradiction. If $ma \leq b$ for all $m \in N$, $m > 0$, then

$$A = \{ma \mid m \in N \text{ and } m > 0\}$$

is a nonempty subset of R, bounded above by b. By the Completeness Axiom there is a least upper bound $u \in R$ for A. Consequently

$$u \geq ma$$

for each $m \in N$, $m > 0$, and therefore

$$u \geq (k + 1)a = ka + a$$

for each $k \in N$, $k > 0$, so that

$$u - a \geq ka$$

for each $k \in N$. Thus

$$ka \leq u - a < u$$

for each $k \in N$, $k > 0$, a contradiction, since $u = $ l.u.b. A. Therefore there is a positive integer n such that $na > b$.

Q.E.D.

Exercises

1. Prove: If m, n are natural numbers and $m < n$, then $m + 1 \leq n$.
2. Prove: If m, n are natural numbers and $m < n + 1$, then $m \leq n$.
3. Let a be a natural number and define

$$N_a = \{x \mid x \in N \text{ and } x \geq a\}.$$

Thus, in particular, $N_0 = N$, the set of all natural numbers. Prove: If (a) K is a nonempty, hereditary subset of N_a and (b) $a \in K$, then $K = N_a$. *Hint:* Use the Well-Ordering Principle. (This form of the PFI is useful on occasion.)

4. Another form of the PFI used in proving many theorems is as follows: Define N_a as in Exercise 3, and let K be a nonempty subset of N_a satisfying the condition:

If $n \in N_a$ and if $\{x \mid a \leq x < n\} \subset K$, then $n \in K$.

Then $K = N_a$.

5. Use the PFI to prove

(a) $1 + 2 + 3 + \cdots + n = \dfrac{n(n+1)}{2}$;

(b) if $z \in N$, $z > 1$, then $\dfrac{z^n - 1}{z - 1} \in N$ for each natural number n;

(c) for each a, z and $n \in N$, $z > 1$,

$$a + az + az^2 + \cdots + az^n = \frac{a(z^{n+1} - 1)}{z - 1};$$

(d) $x_1^2 + x_2^2 + \cdots + x_n^2 \geq 0$ for all natural numbers $x_1, x_2, \ldots, x_n$. Show that $x_1^2 + x_2^2 + \cdots + x_n^2 = 0$ if and only if $x_1 = x_2 = \cdots = x_n = 0$.

6. Prove that *one* is the only natural number having a reciprocal in N.

7. Let $x_1, x_2, \ldots, x_n$ be real numbers. Use the PFI to prove that $|x_1 + x_2 + \cdots + x_n| \leq |x_1| + |x_2| + \cdots + |x_n|$.

15. *The Integers*

The integers are obtained from the natural numbers by augmenting them with the negatives of the natural numbers. Thus if we set

$$-N = \{-x \mid x \in N\},$$

then the set Z of integers is defined as

$$Z = N \cup (-N).$$

As in the case of the natural numbers it is immediate that F.1, F.2, F.3, F.6, 0.1 and 0.2 hold for Z. In addition F.4 is also satisfied by Z. As for F.5, it is easy to check (for the reader) that only 1 and -1 have reciprocals in Z. In discussing the nonnegative elements of Z, we may refer to them as the "nonnegative integers" rather than as the "natural numbers." The elementary properties of Z of greatest interest for us are those based upon the concept of "divisibility."

DEFINITION 38. Let $a, b \in Z$, $b \neq 0$. If there is an integer c such that $a = bc$, then b *divides* a (b is a *divisor* of a; b is a *factor* of a; a is a *multiple* of b). We write "$b \mid a$." If b does not divide a, we write "$b \nmid a$."

Exercises

Prove:
1. If a and b are both nonzero and $b \mid a$, then $|b| \leq |a|$. Hence, if a and b are both positive and $b \mid a$, then $b \leq a$.

2. $b \mid a$ if and only if $-b \mid a$. Hence $b \mid a$ if and only if $|b| \mid |a|$.

3. The only divisors of 1 are ± 1.

4. If $a \mid b$ and $b \mid a$, then $b = \pm a$.

The basic theorem, familiar from elementary arithmetic, is

THEOREM 34.   (Euclid's Division Algorithm). If $a, b \in Z$ and $b \neq 0$, then there exist unique integers q and r such that

(19) $a = qb + r$

where $0 \leq r < |b|$. (q is the *quotient*; r is the *remainder*.)

Proof: We begin by showing that the set

$$S = \{a - x \cdot |b| \mid x \in Z\}$$

contains a nonnegative integer. For, if $a \geq 0$, then

$$a = a - 0 \cdot |b| \in S.$$

On the other hand, if $a < 0$, then $-a > 0$. Since $|b| > 0$ and $b \in Z$, $|b| \geq 1$. Hence $(-a) \cdot |b| \geq -a$, and therefore $a - a \cdot |b| \geq 0$. Since $a - a \cdot |b| \in S$ our assertion is proved in every case.

Next let T be the set of nonnegative elements in S; by the foregoing, $T \neq \emptyset$ and so T contains a smallest integer r. Consequently there is an integer y such that

$$r = a - y \cdot |b|, \ 0 \leq r.$$

We claim that $r < |b|$. For, if $r \geq |b|$, then

$$a - (y + 1) \cdot |b| = (a - y \cdot |b|) - |b| = r - |b| \in T.$$

But $r - |b| < r$; and since r is the smallest integer in T, this is a contradiction.

To prove uniqueness, suppose

$$a = y_1 \cdot |b| + r_1 = y_2 \cdot |b| + r_2$$

where $0 \leq r_1 < |b|$, $0 \leq r_2 < |b|$. Assume $r_1 \leq r_2$. Then $0 \leq r_2 - r_1 < |b|$ and

$$r_2 - r_1 = (y_1 - y_2) \cdot |b|;$$

hence $|b|$ is a divisor of the nonnegative integer $r_2 - r_1$. But if $r_2 - r_1 > 0$, then by Exercise 1, above, $|b| \leq r_2 - r_1$, a contradiction. Therefore $r_2 - r_1 = 0$, $r_1 = r_2$ and $y_1 = y_2$.

Finally, if $b > 0$, then $|b| = b$ and we take $q = y$. If $b < 0$, then $|b| = -b$ and we take $q = -y$. In either case, the integers q and r are such that (19) holds.

Q.E.D.

DEFINITION 39. Let $a, b \in Z$.

(i) An integer d is a *common divisor* of a and b if and only if $d \,|\, a$ and $d \,|\, b$.

(ii) Suppose $a \neq 0$ or $b \neq 0$. An integer $d > 0$ is the *greatest common divisor* (g.c.d.) of a and b if and only if

(a) d is a common divisor of a and b, and

(b) if $c \,|\, a$ and $c \,|\, b$, then $c \,|\, d$.

(iii) a and b are *relatively prime* (or, *coprime*) if and only if g.c.d.$(a, b) = 1$.

If $d =$ g.c.d. (a, b), we also write "$d = (a, b)$" for brevity. Thus, in case a and b are relatively prime, we have $(a, b) = 1$. We shall now prove that every pair of integers, not both zero, has a g.c.d. The first step is a

LEMMA. If $T \neq \emptyset$ is a subset of Z such that

if $x, y \in T$, then $x + y$ and $x - y \in T$,

then T contains a smallest nonnegative integer d and

(20) $$T = \{nd \,|\, n \in Z\}.$$

Further, if $T \neq \{0\}$, then $d > 0$.

Proof: If $T = \{0\}$, then we take $d = 0$.

Suppose $T \neq \{0\}$; we prove, first, that T contains a smallest positive integer.

T contains a positive integer. For $T \neq \{0\}$ implies there is an integer $x \neq 0$ in T. By hypothesis, $0 = x - x \in T$ and $-x = 0 - x \in T$. Since one of $x, -x$ is positive, our assertion follows.

Next let T' be the subset of positive elements in T. $T' \neq \emptyset$ and therefore it contains a smallest integer $d > 0$. The reader should prove

(a) $\{nd \,|\, n \in N\} \subset T$ (*Hint:* use the PFI);

(b) hence $\{nd \,|\, n \in Z\} \subset T$.

We now show that every integer in T is a multiple of d. This result, together with (b), yields (20).

Suppose $y \in T$ is not a multiple of d. Then there are integers q and r such that

$$y = qd + r, \; 0 \leq r < |d| = d.$$

Now $r > 0$, since otherwise $y = qd$. But then

$$r = y - qd$$

and so $r \in T$. Thus $0 < r < d$ and $r \in T$, contrary to the choice of d as the smallest positive integer in T. Hence every element in T is a multiple of d, and the Lemma is proved.

THEOREM 35. If a and b are integers not both zero, then there exist integers s and t such that

$$\text{g.c.d. } (a,b) = sa + tb.$$

Proof: Define $T = \{xa + yb \mid x, y \in Z\}$; $T \neq \emptyset$ and if $u, v \in T$ then $u + v$ and $u - v$ are in T. Hence T contains a smallest integer $d > 0$ such that $T = \{nd \mid n \in Z\}$. By definition of T, there exist integers s and t such that

$$d = sa + tb.$$

We claim that $d = $ g.c.d. (a, b).

Since $a = 1 \cdot a + 0 \cdot b \in T$ and $b = 0 \cdot a + 1 \cdot b \in T$, by the Lemma, $d \mid a$ and $d \mid b$. Suppose $c \mid a$ and $c \mid b$; we prove that $c \mid d$ and therefore (since $d > 0$) $d = $ g.c.d. (a, b). From $c \mid a$ and $c \mid b$ we have $a = cu$, $b = cv$ where $u, v \in Z$. Hence

$$d = sa + tb$$
$$= s(cu) + t(cv)$$
$$= (su + tv)c$$

and therefore $c \mid d$.

Q.E.D.

The computation of g.c.d.'s is carried out by repeated application of the division algorithm. Details are given in the exercises below. *Note:* If $a = 0$ and $b \neq 0$, then g.c.d. $(a, b) = |b|$; similarly, if $a \neq 0$, $b = 0$, then g.c.d. $(a, b) = |a|$. In the following exercises it is assumed that a and b are both nonzero integers.

Exercises

1. Prove: g.c.d. $(a, b) = $ g.c.d. $(-a, b) = $ g.c.d. $(a, -b) = $ g.c.d. $(-a, -b)$.

2. Assume a and b are both positive integers and $a \nmid b$ and $b \nmid a$. Define $q_i, r_i \in Z$ by

$$a = q_1 b + r_1, \, 0 \leq r_1 < b,$$
$$b = q_2 r_1 + r_2, \, 0 \leq r_2 < r_1,$$
$$r_i = q_{i+2} r_{i+1} + r_{i+2}, \, i \in N, \, i \geq 1.$$

Prove there exists a positive integer n such that $r_{n+1} = 0$.

3. By Exercise 2 there is a smallest positive integer m such that $r_{m+1} = 0$. Prove that g.c.d. $(a, b) = $ g.c.d. $(b, r_1) = $ g.c.d. $(r_i, r_{i+1}), 1 \leq i \leq m - 1$, hence that $r_m = $ g.c.d. (a, b).

4. Use the method of the foregoing exercises to find g.c.d. $(991, 236)$.

DEFINITION 40. An integer $p \neq 0$, ± 1 is *prime* if and only if $d \mid p$ implies $d = \pm 1$ or $d = \pm p$.

The reader should verify that 2, 3 and 5 are primes.

THEOREM 36. If p is a prime and $p \mid ab$, then $p \mid a$ or $p \mid b$.

> *Proof:* We assume $p \nmid a$ and deduce $p \mid b$. Since $p \nmid a$ and p is a prime, g.c.d. $(p, a) = 1$. By Theorem 35 there exist integers s and t such that
>
> $$1 = sp + ta.$$
>
> Then
>
> $$b = spb + tab.$$
>
> Since $p \mid ab$, there is an integer c such that $ab = cp$. Hence
>
> $$b = spb + tab = (sb + tc)p$$
>
> where $sb + tc \in Z$. Therefore $p \mid b$.
>
> Q.E.D.

The reader should prove the following corollaries.

COROLLARY 1. If p is prime and $p \mid a_1 a_2 \ldots a_n$, $n \geq 1$, then $p \mid a_i$ for some i, $1 \leq i \leq n$.

COROLLARY 2. If $c \mid ab$, and a, c are relatively prime, then $c \mid b$.

We now state an important result, the "Unique Factorization Theorem" or the "Fundamental Theorem of Arithmetic." The proof will not be given here since a generalization is stated and proved in Chapter 4, pages 186–189.

THEOREM 37.   (i) If a is an integer $\neq 0$, ± 1, then there is a positive integer $r \geq 1$ and positive primes $p_1, p_2, \ldots, p_r$, such that

$$a = \pm 1 \cdot p_1 \cdot p_2 \cdot \ldots \cdot p_r.$$

(ii) If $p_1 \cdot p_2 \cdot \ldots \cdot p_r = q_1 \cdot q_2 \cdot \ldots \cdot q_s$ where the p's and q's are positive primes, then $r = s$, each $p_i =$ some q_j, and each $q_j =$ some p_i.

(i) and (ii) are summed up by: Every integer $a \neq 0$, ± 1, is ± 1 times a product of positive primes which is unique to within order of the primes.

Exercises

1. Let $d =$ g.c.d. (a, b), where a and b are nonzero and where $a = xd$ and $b = yd$. Prove that g.c.d. $(x, y) = 1$.
2. If $a \neq 0$, prove that g.c.d. $(ab, ac) = |a|$ g.c.d. (b, c).

3. If a,b,c are nonzero integers, define
$$\text{g.c.d. } (a, b, c) = \text{g.c.d. } (\text{g.c.d. } (a, b), c).$$
Prove: g.c.d. $(a, b, c) = $ g.c.d. $(\text{g.c.d. } (a,c), b) = $ g.c.d. $(\text{g.c.d. } (b, c), a)$.

4. If a, b, c are nonzero integers, prove that there exist integers r, s, t such that g.c.d. $(a, b, c) = ra + sb + tc$.

5. Generalize the definition of g.c.d. for the case of n integers, $n \geq 2$. Prove: If $a_1, a_2, \ldots, a_n, n \geq 2$, are nonzero integers, then there exist integers $r_1, r_2, \ldots, r_n$ such that g.c.d. $(a_1, a_2, \ldots, a_n) = r_1 a_1 + r_2 a_2 + \ldots + r_n a_n$.

6. Let a and b be nonzero integers. An integer $m > 0$ is a *least common multiple* (l.c.m.) of a and b if and only if
   (i) $a \mid m$ and $b \mid m$; and
   (ii) if $a \mid n$ and $b \mid n$, then $m \mid n$.
Prove that $|ab| = (\text{g.c.d. } (a, b)) \cdot (\text{l.c.m. } (a, b))$.

7. An integer n is *even* if and only if there is an integer q such that $n = 2q$; otherwise it is *odd*. Prove: (a) If n is odd, then there is a $q \in Z$ such that $n = 2q + 1$; (b) $m + n$ is even if and only if m and n are both even or both odd; (c) mn is even if and only if at least one of m and n is even.

16. *The Rational Numbers*

DEFINITION 41. The set of *rational numbers* (or, *fractions*), Q, is the subset of R defined by

$$Q = \left\{ \frac{a}{b} \,\middle|\, a, b \in Z \text{ and } b \neq 0 \right\}.$$

If $a/b \in Q$, then a is the *numerator* and b is the *denominator* of the given fraction.

Since $a/b \cdot c/d = ac/bd$ and $a/b + c/d = (ad + bc)/bd$, it follows that sums and products of rational numbers are again rationals. We have seen that if a, b, c, d are real numbers, $b \neq 0$, $d \neq 0$, then $a/b = c/d$ if and only if $ad = bc$. This same result holds if a/b and c/d are rationals. The reader can check, without difficulty, that F.1–F.6, 0.1 and 0.2 are all valid for Q. Also the rational numbers Q contain the integers Z as a subset. For, since $a = a/1$ for each $a \in Z$, it follows that Z has been included in Q by Definition 41.

Exercises

1. Let $a/b \in Q$; prove (a) $a/b > 0$ if and only if $ab > 0$. (b) a/b is negative if and only if $ab < 0$.

2. Let m and n be positive integers such that $m < n$, and let $r \in Q$. Prove (a) If $1 < r$, then $r^m < r^n$. (b) If $0 < r < 1$, then $r^n < r^m$.

We now prove that there is no square root of 2 in Q. Hence Q is not complete in the sense of our axiom and $Q \subset R$, but $Q \neq R$ as remarked on page 35, footnote.

THEOREM 38.   There is no rational number whose square is 2.

> *Proof:* Suppose p/q is a rational number such that p and q are relatively prime and $(p/q)^2 = 2$. Then $p^2 = 2q^2$, i.e., p^2 is even and therefore (Exercise 7, (c), page 47) p is even, say, $p = 2r$. Thus $2q^2 = p^2 = 4r^2$ whence $q^2 = 2r^2$. But $2r^2$ is even; therefore, so is q^2, and q is also even. Consequently, p and q are both even, contrary to the assumption that g.c.d. $(p, q) = 1$.
>
> Q.E.D.

17. *The Complex Numbers*

The final number system to be studied in this chapter is the system of *complex numbers*, C. The previous structures N, Z and Q have all been investigated as systems contained within R. Our study of C will proceed in an entirely different way. Instead of considering C as contained within some previously given structure, we shall *build* C out of R.

Let $C = R \times R$ and define binary operations $\oplus$ and $\odot$ on C as follows:

$$(a, b) \oplus (c, d) = (a + c, b + d)$$

and

$$(a, b) \odot (c, d) = (ac - bd, ad + bc).$$

Note that the operations $+$, $\cdot$ indicated in the above two equations are the ordinary sum and product of real numbers. The $\oplus$ and $\odot$ on the left are new binary operations defined by the two equations.

> DEFINITION 42.   The elements of C are *complex numbers*. C, together with the binary operations $\oplus$ and $\odot$, is the *complex number field*.

We now prove that the complex number field satisfies all the field axioms F.1–F.6. Thus we verify:

> F.1.′   For each (a, b) and (c, d) in C, $(a, b) \oplus (c, d) = (c, d)$ $\oplus (a, b)$ and $(a, b) \odot (c, d) = (c, d) \odot (a, b)$.

> *Proof:* $(a, b) \oplus (c, d) = (a+c, b+d) = (c+a, d+b) = (c, d) \oplus (a, b)$.
> $(a, b) \odot (c, d) = (ac-bd, ad+bc) = (ca-db, da+cb)$
> $= (c, d) \odot (a, b)$.

F.2.′ If (a, b), (c, d), (e, f) are complex numbers, then

$$(a, b) \oplus ((c, d) \oplus (e, f)) = ((a, b) \oplus (c, d)) \oplus (e, f)$$

and

$$(a, b) \odot ((c, d) \odot (e, f)) = ((a, b) \odot (c, d)) \odot (e, f).$$

Proof: Exercise.

F.3.′ There exist unique complex numbers $(0, 0)$ and $(1, 0)$ such that for each $(a, b) \in C$,

$$(0, 0) \oplus (a, b) = (a, b) \oplus (0, 0) = (a, b)$$

and

$$(1, 0) \odot (a, b) = (a, b) \odot (1, 0) = (a, b).$$

Proof: $(0, 0) \oplus (a, b) = (0+a, 0+b) = (a, b) = (a+0, b+0) = (a, b)$ $\oplus (0, 0)$.
   $(1, 0) \odot (a, b) = (1 \cdot a - 0 \cdot b, 1 \cdot b + 0 \cdot a) = (a, b) = (a \cdot 1 - b \cdot 0, b \cdot 1 + a \cdot 0) = (a, b) \odot (1, 0)$.
The uniqueness proof is left to the reader.

F.4.′ For each $(a, b) \in C$ there is a unique $(c, d) \in C$ such that

$$(a, b) \oplus (c, d) = (c, d) \oplus (a, b) = (0, 0).$$

Proof: Take $(c, d) = (-a, -b)$. The uniqueness is trivial.

F.5.′ For each $(a, b) \in C$, $(a, b) \neq (0, 0)$, there is a unique $(c, d) \in C$ such that

$$(a, b) \odot (c, d) = (c, d) \odot (a, b) = (1, 0).$$

Proof: Take

$$(c, d) = \left(\frac{a}{a^2 + b^2}, \frac{-b}{a^2 + b^2} \right).$$

Again, the uniqueness is easy to verify.

F.6.′ For each (a, b), (c, d) and (e, f) in C,
   $(a, b) \odot ((c, d) \oplus (e, f)) = (a, b) \odot (c, d) \oplus (a, b) \odot (e, f).$

Proof: Exercise.

In Theorem 21 we proved that the square of every nonzero number in R is positive. From this it follows that the negative real numbers do not have square roots in R. The complex numbers, by contrast, have the following property:

THEOREM 39. Each element $(a, b) \in C$, $(a, b) \neq (0, 0)$, has two square roots in C. In other words, there exist (x, y) and (u, v) in C, $(x, y) \neq (u, v)$ such that

$$(x, y)^2 = (x, y) \odot (x, y) = (a, b)$$

and

$$(u, v)^2 = (u, v) \odot (u, v) = (a, b).$$

Proof: It is easy to verify that each of the complex numbers

$$\left(\sqrt{\frac{a + \sqrt{a^2 + b^2}}{2}}, \; \frac{b}{2\sqrt{\frac{a + \sqrt{a^2 + b^2}}{2}}} \right)$$

and

$$\left(-\sqrt{\frac{a + \sqrt{a^2 + b^2}}{2}}, \; \frac{-b}{2\sqrt{\frac{a + \sqrt{a^2 + b^2}}{2}}} \right)$$

is a square root of (a, b).

Q.E.D.

A second major difference between C and R is that the complex numbers *cannot* be ordered in a way consistent with addition and multiplication (see page 34). Thus, there is no way of selecting a nonempty subset P of C such that

(i) for each $(a, b) \in C$, exactly one of $(a, b) \in P$, $(a, b) = (0, 0)$ or $(-a, -b) \in P$ holds;

(ii) if $(a, b), (c, d) \in P$, then $(a, b) \oplus (c, d) \in P$ and $(a, b) \odot (c, d) \in P$.

Indeed, suppose there were such a subset P. By the same proof as in Theorem 21 it follows that for each $(a, b) \in C$, $(a, b)^2 \in P$. In particular, $(1, 0)^2 \in P$. But $(1, 0)^2 = (1, 0)$; hence $(1, 0) \in P$. By (i) above, $(-1, 0) \notin P$. Now consider $(0, 1)$; by (i) exactly one of $(0, 1) \in P$ or $(0, -1) \in P$ holds. If $(0, 1) \in P$, then by (ii)

$$(0, 1)^2 = (-1, 0) \in P,$$

a contradiction. On the other hand, if $(0, -1) \in P$, then by (ii)

$$(0, -1)^2 = (-1, 0) \in P,$$

the same contradiction. Thus our assumption yields a contradiction.

Although the complex numbers differ in essential ways from the reals, C does contain a carbon copy[5] of R. Thus the subset

$$R' = \{(a, b) \,|\, (a, b) \in C \text{ and } b = 0\}$$

is easily seen to satisfy not only F.1–F.6, but also 0.1 and 0.2 as well as the Completeness Axiom.

The notation we adopted to construct the complex numbers is, of course, not the familiar one. The standard notation is introduced as follows:

First, note that $(a, b) = (a, 0) \oplus (0, b)$ and $(0, b) = (b, 0) \odot (0, 1)$.

---

[5] That is, an *isomorphic* copy. See Hamilton and Landin, *Set Theory and the Structure of Arithmetic*, p. 255, Boston (Allyn and Bacon), 1961.

Thus

(21) $$(a, b) = (a, 0) \oplus (b, 0) \odot (0, 1).$$

Henceforth we make the following substitutions:

in place of "$(x, 0)$" write "x";
in place of "$(0, 1)$" write "i";
in place of "$\oplus$" write "$+$";
in place of "$\odot$" write "$\cdot$".

Then (21) becomes

(22) $$(a, b) = a + b \cdot i.$$

Seen in this way, the real numbers are simply the complex numbers with second component zero.

Exercises

1. State the rules for computation with complex numbers in the form (22). Verify that $i^2 = -1$.

2. What, if anything, is wrong with $1 = \sqrt{1} = \sqrt{(-1)(-1)} = \sqrt{-1}\sqrt{-1} = -1$?

2

The Theory of Groups

1. *The Group Concept*

A great deal of abstract algebra is concerned with the study of structures such as groups, rings, integral domains, fields, vector spaces, modules and so on. Among these structures, groups are undoubtedly the easiest to define. The reason is that whereas groups involve (at least, initially) but *one* binary operation, the remaining structures all involve at least *two*. It is therefore appropriate to begin the study of abstract algebra with group theory. One might add that group theory is one of the most universal workhorses in the mathematician's stable; not only is this theory used extensively in the development of abstract algebra itself, but also it has important applications in a wide variety of disciplines including geometry, analysis, topology, mathematical physics and engineering.

We initiate the study of groups by considering a few examples.

EXAMPLE 1. Let Z be the set of integers. The structure consisting of Z and the binary operation $+$ has the following properties:

(i) if $a, b, c \in Z$, then $a + (b + c) = (a + b) + c$, i.e., the associative law holds;

(ii) there is an integer, 0, such that for each $a \in Z$, $0 + a = a$;

(iii) for each integer a there is an integer b such that $b + a = 0$; b is $-a$.

The structure consisting of the set Z together with the binary operation $+$ is a *group*.

EXAMPLE 2. Let Q'_+ be the set of positive rational numbers with multiplication as the binary operation. It may be observed that

(i) if $a, b, c \in Q'_+$, then $a(bc) = (ab)c$, i.e., the associative law holds;

(ii) there is a rational number, 1, in Q'_r such that, for each $a \in Q'_r$, $1 \cdot a = a$;

(iii) for each $a \in Q'_r$, there is a rational number $b \in Q'_r$ such that $b \cdot a = 1$.

The rational number b is the reciprocal of a. The structure Q'_r is a *group*.

Examples 1 and 2 are numerical in character and introduce nothing especially new. Our next two examples of groups are not numerical.

EXAMPLE 3. (See Figure 6) For the purposes of this example, think of a piece of cardboard cut in the shape of an equilateral triangle and placed on a sheet of paper with its vertices at the points A, B and C on the sheet of paper.

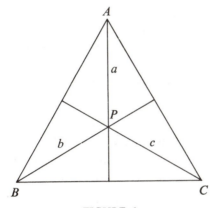

FIGURE 6

The altitudes a, b and c are considered as lying on the sheet of paper beneath the cardboard triangle. A *rigid motion* or, more simply, a *motion* of the triangle is any displacement throughout which the shape of the triangle is unaltered. The question is: What are the motions of the triangle which—after the motions are carried out—leave the vertices at the points A, B and C. Let

R be the counterclockwise rotation of the triangle, about P, through $120°$;

T be the counterclockwise rotation of the triangle, about P, through $240°$;

F_a be the motion which flips the triangle, exactly once, over the altitude a;

F_b and F_c be defined similarly with respect to the altitudes b and c.

Any motion which leaves the triangle exactly as it was before the motion was applied is denoted by "I." If X and Y are motions, then $X * Y$ is the motion obtained by first performing the motion X followed by the motion Y. For instance,

$$R * F_a = F_b.$$

The set with which we are concerned is

$$M = \{I, R, T, F_a, F_b, F_c\}.$$

From the definition of $*$ one can verify that $*$ is a binary operation on M and that if $X, Y, Z \in M$, then $X * (Y * Z) = (X * Y) * Z$. It would be a long and tedious job to verify that the associative law holds in all cases. (Indeed, how many verifications are required?) We therefore suggest that the reader do only a sufficient number of them to convince himself that $*$ is a binary, associative operation on M. Also, note that $T = R * R$ and that

$$R * T = T * R = I, \; F_a * F_a = I, \; F_b * F_b = I, \; F_c * F_c = I, \; I * I = I.$$

To sum up, we see that

(i) if X, Y, Z are motions in M, then $X * (Y * Z) = (X * Y) * Z$;
(ii) there is a motion I such that for each $X \in M$, $I * X = X$;
(iii) for each $X \in M$, there is a $Y \in M$ such that $Y * X = I$.

The structure consisting of M and $*$ is a *group*.

EXAMPLE 4.   Let $T = \{a, b, c\}$ be a set consisting of three elements. Consider all bijections $T \longrightarrow T$; there are six such mappings, namely,

$$\psi_1: \quad \psi_1(a) = a, \; \psi_1(b) = b, \; \psi_1(c) = c;$$
$$\psi_2: \quad \psi_2(a) = b, \; \psi_2(b) = c, \; \psi_2(c) = a;$$
$$\psi_3: \quad \psi_3(a) = c, \; \psi_3(b) = a, \; \psi_3(c) = b;$$
$$\psi_4: \quad \psi_4(a) = b, \; \psi_4(b) = a, \; \psi_4(c) = c;$$
$$\psi_5: \quad \psi_5(a) = c, \; \psi_5(b) = b, \; \psi_5(c) = a;$$
$$\psi_6: \quad \psi_6(a) = a, \; \psi_6(b) = c, \; \psi_6(c) = b.$$

The set from which we construct a group is the set

$$S = \{\psi_1, \psi_2, \psi_3, \psi_4, \psi_5, \psi_6\}.$$

The binary operation on S used to construct our group is the composition of mappings. For instance,

$$\psi_3 \circ \psi_2 = \psi_1,$$

as can be determined by simple computation. Again, one can verify that

(i) if ψ_i, ψ_j, ψ_k are elements in S, then $\psi_i \circ (\psi_j \circ \psi_k) = (\psi_i \circ \psi_j) \circ \psi_k$;
(ii) there is an element, ψ_1, in S, such that for each $\psi_i \in S$, $\psi_1 \circ \psi_i = \psi_i$;
(iii) for each $\psi_i \in S$, there is a $\psi_j \in S$ such that $\psi_j \circ \psi_i = \psi_1$.

The structure consisting of S and the binary operation $\circ$ is a *group*.

Whatever else one may say about our four examples, this much is clear: *They all share properties* (i), (ii) and (iii). It is true that they have many other properties in common, but a large number of the shared properties are deducible from the three we have chosen to identify. Indeed, all structures consisting of a set together with a binary operation satisfying (i), (ii) and (iii) have in common a highly developed theory.

DEFINITION 1.   A *group* is a structure consisting of a set G and a binary operation $*$ on G such that

G.1.   if $a, b, c \in G$, then $a * (b * c) = (a * b) * c$; i.e., the binary operation is associative;

G.2.   there is an element $e \in G$ such that for each $a \in G$, $e * a = a$; e is a *left neutral element* for G;

G.3.   for each $a \in G$, there is an element $b \in G$ such that $b * a = e$; b is a *left inverse* of a with respect to e.

The binary operation $*$ on G is the *group operation*.

If the structure consisting of the set G and the binary operation $*$ on G is a group, then G.2 implies that the set G is nonempty.

Before proceeding with the theory, let us emphasize that *not* every simple structure possessing a binary operation is a group.

EXAMPLE 5.   Let $G = Z$, the set of integers, and let the binary operation be multiplication. As we know from elementary arithmetic, the multiplication of integers is associative, and 1 is a left neutral element. However, the element $5 \in Z$ has no inverse *in Z*. To be sure, $1/5 \cdot 5 = 1$ but $1/5$ is *not* an element in Z. To violate the definition of group, it is sufficient that there be *one* element in Z having no left inverse in Z. Thus G.3 does not hold and therefore the system consisting of Z and the binary operation of multiplication is not a group. We note, in passing, that no element of Z, other than 1 and -1, has a left inverse in Z.

EXAMPLE 6.   Let $G = Z$ be the set of all integers and define a binary operation $*$ on Z by

$$a * b = a$$

for each a and each $b \in Z$. In this example it is easy to see that the operation $*$ is associative, but clearly there is no left neutral element. Hence the structure consisting of Z and $*$ is not a group.

EXAMPLE 7.   Let $G = \{a, b, c, d\}$ be a set consisting of four elements and let the binary operation $*$ be defined by the following table:

TABLE 2

$*$	a	b	c	d
a	a	b	c	d
b	b	a	c	d
c	c	d	a	b
d	d	a	c	a

Clearly a is a left (and also, a right) neutral element. For each $x \in G$, x is its own left inverse. However, since

$$c * (b * d) = c * d = b$$

and

$$(c * b) * d = d * d = a,$$

the associative law does not hold, and the structure we have defined is not a group.

Exercises

1. Let Q be the set of rational numbers and let $G = Q - \{1\}$. Define a binary operation $*$ on G by means of

$$a * b = a + b - ab$$

for all $a, b \in G$. Show that the structure so defined is a group.

2. Let Q be the set of rational numbers and let $G = Q - \{-1\}$. Define a binary operation $*$ on G by means of

$$a * b = a + b + ab$$

for all $a, b \in G$. Verify that this structure is a group.

3. Again let Q be the set of rational numbers. Is Q a group with respect to the operation of multiplication? Is Q a group with respect to the operation of addition? Is $Q - \{0\}$ a group with respect to multiplication?

4. Let Z be the set of integers and let $G = Z \times Z$. Define an operation $*$ on Z by means of

$$(a, b) * (c, d) = (a + c, b + d)$$

for all $(a, b), (c, d) \in Z \times Z$. Verify that this structure is a group.

5. Construct a group table for the group of Example 3; for the group of Example 4.

6. (See Figure 7) Consider a piece of cardboard cut in the shape of a square and placed on a sheet of paper with its vertices at the points A, B, C and D on the sheet of paper. Let F_x be the motion of flipping the square, exactly once, over the horizontal line x, let F_y be the motion of flipping the square, exactly once,

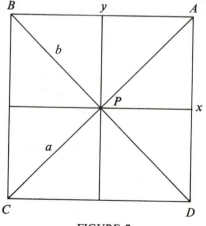

FIGURE 7

over the vertical line y, let S be the rotation of the square through 180°, counterclockwise about P, and let I be any motion which leaves the vertices of the square exactly as they were before the motion started. Define the binary operation $*$ as in Example 3. Verify[1] that the set $\{I, S, F_x, F_y\}$ with the binary operation $*$ is a group.

7. With the same gadgets as in Exercise 6, let F_a be the motion of flipping the square, exactly once, over the diagonal a, and let F_b be defined similarly with respect to the diagonal b. Further, let R and T be counterclockwise rotations of the square, about P, through 90° and 270° respectively. With the binary operation $*$ defined as in Exercise 6, verify[1] that the set $\{I, R, S, T, F_x, F_y, F_a, F_b\}$ is a group with the binary operation $*$.

A few remarks on group notation and terminology will be useful. In writing of a group, we have referred to a structure consisting of a set together with a binary operation on that set. It should be emphasized that the symbols G and $*$ are variables whose ranges are, respectively, certain collections of sets and certain collections of binary operations. As in other branches of mathematics, we shall need many different variables. Thus, in addition to the symbol G, we shall use symbols such as H, K, M, N, etc., as variables having the same range as G, and symbols such as $*$, $\circ$, $+$, and $\cdot$ (or, juxtaposition) as variables whose range is that of $*$. The use of $+$ and $\cdot$ (juxtaposition) requires a word of caution; the following example illustrates a difficulty that arises:

For legal purposes, the expression "John Doe" is used as a variable whose range is some specified set of men. Thus one reads of the issuance of John Doe warrants in criminal cases. When sufficient evidence has been accumulated against some man, then in effect, the expression "John Doe" is replaced by the name of the man to be charged. Now, it might actually happen that the name of the person to be charged is "John Doe." In such an unlikely event, one might—in the absence of detailed information—confuse the variable *John Doe* with the name "John Doe."

A similar situation arises in group theory. On the one hand, the symbol "$+$" is commonly used as a variable whose range is a set of binary operations. This is its use in the sentence:

(†) Let G be a group with group operation $+$.

On the other hand, the "$+$" is also used to designate specific binary operations. Thus in the group of integers (Example 1), the group operation is denoted by "$+$." Whether the "$+$" occurs as a variable or as a designation of a specific binary operation, one customarily speaks of "additive groups." For instance, the sentence (†) might be written

Let G be an additive group.

And one might also speak of "the additive group of the integers."

---

[1] The number of verifications required to show that the operations defined in Exercises 6 and 7 are binary and associative is extremely long. Do about five or six to convince yourself of the plausibility of the associative law.

The symbol "·" (juxtaposition) is also in common use both as a variable and as a denotation for specific binary operations. If one writes

(††)   Let G be a group with group operation·,

then the "·" is a variable. But in Example 2, the "·" denotes ordinary multiplication of rational numbers. Whether the "·" occurs as a variable or as a designation of a specific binary operation, one refers to a "multiplicative group." For instance, (††) is also written

Let G be a multiplicative group,

and one refers (Example 2) to "the multiplicative group of the positive rationals."

Wherever it is convenient and leads to no confusion, we shall use the symbols "·" (juxtaposition) and "+" to denote binary operations on groups. Many of our groups will therefore be termed "multiplicative" or "additive." If the notation for the group operation is not specified, we shall generally use multiplicative notation.

We conclude this section by observing that from the definition of a group one may deduce a *generalized associative law*. The technical details are of minor interest at this point, and therefore we illustrate with examples. Suppose G is a group with group operation $*$, and a, b, c, d are elements in G. Then

$$((a * b) * c) * d = (a * b) * (c * d)$$
$$= a * (b * (c * d))$$
$$= a * ((b * c) * d)$$
$$= \text{etc.}$$

Similarly, if a, b, c, d, e are elements in G, then

$$(((a * b) * c) * d) * e = ((a * b) * c) * (d * e)$$
$$= (a * b) * (c * (d * e))$$
$$= \text{etc.}$$

In effect, the generalized associative law states that in a product of a finite number of group elements, it is immaterial how the parentheses are inserted provided, of course, they are inserted in a meaningful way.

2.   Some Simple Consequences of the Definition of Group

DEFINITION 2.   Let G be a group. An element $f \in G$ is a *right neutral element* for G if and only if $a \cdot f = a$ for each $a \in G$. An element $f \in G$ is a *neutral element* for G if and only if it is both a left and a right neutral element.

A casual examination of the groups described in Examples 1-4 reveals that in each group the specified left neutral element is, in fact, a neutral element, and moreover that each group has only one neutral element. These facts were not listed as part of the definition of group for the simple reason that they are deducible as consequences of G.1—G.3.

THEOREM 1. Each left and each right neutral element of a group G is a neutral element. Moreover, a group has exactly one neutral element.

Proof: By G.2 and G.3 there is a left neutral element $e \in G$, and for each $a \in G$ there exists an element b in G for which $ba = e$. We prove, first, that this particular left neutral element, e, is also a neutral element for G. By Definition 2 it suffices to show that

$$ae = a$$

for each $a \in G$.

Now if b is a left inverse of a with respect to e, and c is a left inverse of b with respect to e, then

$ae = e(ae)$	(e is a left neutral element)
$\quad = (ea)e$	(associative law)
$\quad = ((cb)a)e$	($cb = e$)
$\quad = (c(ba))e$	(associative law)
$\quad = (ce)e$	($ba = e$)
$\quad = c(ee)$	(associative law)
$\quad = ce$	(e is a left neutral element)
$\quad = c(ba)$	($ba = e$)
$\quad = (cb)a$	(associative law)
$\quad = ea$	($cb = e$)
$\quad = a$	(e is a left neutral element).

Therefore, e is right neutral, hence it is a neutral element.

Next let f be a left neutral element for G. Then $fe = e$; on the other hand, we have just proved that e is a neutral element so that $fe = f$. Therefore, $e = f$ and so e is the only left neutral element in G. Finally, if g is a right neutral element it follows as above that $e = g$ (details?), hence e is the only neutral element in G.

Q.E.D.

By reason of Theorem 1 we may speak of "*the* neutral element" of a group. Moreover, we had defined the term "left inverse" with respect to a particular neutral element e. Theorem 1 shows that the expression "with respect to the left neutral element e" may be deleted in discussing left inverses.

DEFINITION 3. Let G be a group with neutral element e and let $a \in G$. An element $b \in G$ such that $ab = e$ is a *right inverse* of a; b is an *inverse* of a if and only if b is both a left inverse and a right inverse of a.

Examples 1-4 suggest the conjecture that every left inverse is an inverse, and that each element in a group has exactly one inverse. These results are deducible from Definition 1 and Theorem 1, as follows:

THEOREM 2. Let G be a group with neutral element e. Then each $a \in G$ has one and only one left inverse in G, and one and only one right inverse in G. Further, these inverses coincide.

Proof: Let b be a left inverse of a. We prove: (i) b is also a right inverse of a, hence b is an inverse of a, and (ii) if c is a left inverse of a, then $c = b$.

(i) By G. 3 we know that b has a left inverse, say, d. Thus $db = e$ and

$$
\begin{aligned}
ab &= e(ab) && \text{(e is the neutral element for G)} \\
   &= (db)(ab) && \text{(since $db = e$)} \\
   &= d((ba)b) && \text{(by G.1)} \\
   &= d(eb) && \text{(since $ba = e$)} \\
   &= db \\
   &= e.
\end{aligned}
$$

Therefore, by Definition 3, b is a right inverse of a, hence b is an inverse of a.

$$
\begin{aligned}
\text{(ii)} \quad c &= ce && \text{(e is the neutral element of G)} \\
   &= c(ab) && \text{($ab = e$)} \\
   &= (ca)b && \text{(G.1)} \\
   &= eb && \text{($ca = e$)} \\
   &= b.
\end{aligned}
$$

The statements corresponding to (i) and (ii) with "left" and "right" interchanged are proved similarly.

Q.E.D.

Because of Theorem 2, we speak of "*the* inverse of an element" in a group G. In a multiplicative group the neutral element will usually be denoted by "e" and the inverse of an element a will be written "a^{-1}." Thus, for each element a in a multiplicative group G, we have

$$a^{-1}a = aa^{-1} = e.$$

In an additive group we shall usually denote the neutral element by "0" (zero) and the inverse of an element a by "$-a$." For each element a in an additive group, we have

$$(-a) + a = a + (-a) = 0.$$

Suppose that G is a group with respect to the binary operation $*$. If a and b are elements in G, then there are elements x and y in G such that

(1)
$$a * x = b,$$

and

(2)
$$y * a = b;$$

indeed,

$$x = a^{-1} * b \quad \text{and} \quad y = b * a^{-1}$$

do the trick. We now prove that equations (1) and (2) provide a characterization of groups; this characterization is occasionally useful as an alternative to Definition 1.

THEOREM 3. Let $*$ be an associative, binary operation on a nonempty set G. Then G is a group with respect to $*$ if and only if for each a and each b in G there exist x and y in G such that equations (1) and (2) hold.

Proof: First, let $*$ be an associative, binary operation on G, $G \neq \emptyset$, such that equations (1) and (2) hold. To prove that G is a group we verify conditions G.2 and G.3 of Definition 1.

Let $a \in G$. By equation (2) there is an element $e \in G$ such that $e * a = a$. Now if $b \in G$, then by (1) there is an $x \in G$ such that $a * x = b$. Hence

$$e * b = e * (a * x) = (e * a) * x = a * x = b;$$

thus, for each $b \in G$, $e * b = b$ and so e is a left neutral element for G. Hence G.2 is verified.

To check G.3 observe that by (2), for each $a \in G$, there is an element $y \in G$ such that $y * a = e$; the element y is a left inverse of a.

The proof of the converse has already been given in the remarks preceding the statement of Theorem 3.

Q.E.D.

Exercises

1. Rewrite the proofs of Theorems 1 and 2 for additive groups; rewrite the proof of Theorem 3 for multiplicative and for additive groups.

2. Let G be a set, and let $*$ be a binary operation on G such that G.1 holds. Suppose that G.2 and G.3 are replaced by the following conditions:

G.2' There is an element $f \in G$ such that for each $a \in G$, $a * f = a$, i.e., f is a right neutral element; and

G.3′ For each $a \in G$, there is an element $b \in G$ such that $a * b = f$; b is a right inverse of a with respect to f.

Outline a proof that the two sets of conditions, G.1, G.2 and G.3, and G.1, G.2′ and G.3′, imply each other, hence determine the same concept of group.

3. Use Example 6 to show that a structure satisfying G.1, G.2′ and G.3 need not be a group.

4. Prove: If x is an element in a group G, then $(x^{-1})^{-1} = x$. What is the corresponding equation if G is an additive group?

Another quick examination of Examples 1-4 shows that in Examples 1 and 2, the binary operations $+$ and $\cdot$, respectively, are commutative. On the other hand, in Example 3,

$$F_a * R = F_b \qquad \text{and} \qquad R * F_a = F_c,$$

so that $F_a * R \neq R * F_a$, and therefore the binary operation $*$ in Example 3 is noncommutative. Similarly, in Example 4,

$$\psi_3 \circ \psi_4 = \psi_6 \qquad \text{and} \qquad \psi_4 \circ \psi_3 = \psi_5,$$

so that the operation of composition is noncommutative. These examples indicate that groups may be separated into two classes in accordance with

DEFINITION 4.   A group G such that

G.4 $ab = ba$ for each a and each b in G, is an *Abelian* (named after the Norwegian mathematician Nils Henrik Abel) or *commutative* group. If there exist x and y in G such that $xy \neq yx$, the group is *non-Abelian* or *noncommutative*.

In this text, as in most others, the "$+$" notation is used only for Abelian groups. If any other notation is used for the group operation, it will be specified whether or not the group is Abelian in case the information is pertinent.

Note that if G is Abelian the proofs of Theorems 1 and 2 are immediate.

Exercises

1. Find the inverse of each element in the group described in Example 4.

2. Let a, x and y be elements in a group G. Prove: $ax = ay$ if and only if $x = y$.

3. Let a and b be elements in a group G. Prove that $(ab)^{-1} = b^{-1}a^{-1}$. (*Hint:* Use the fact that each element in G has exactly one inverse.) Write the corresponding equation in a group in which the group operation is written additively.

4. Let G be a group. Prove: If x is an element in G such that for some $y \in G$, $xy = y$, then x is the neutral element for G. Write out the proof for additive groups.

5. Let G be a (multiplicative) group such that for each $x \in G$, $xx = e$. Prove that G is Abelian.

6. Prove that every group containing five or fewer elements is Abelian.

3. Powers of Elements in a Group

In the following we shall have to refer to *powers* of elements in a group. Suppose G is a group and x is an element in G. As in elementary arithmetic we set

$$x^2 = x \cdot x,$$
$$x^3 = x^2 \cdot x = (x \cdot x) \cdot x,$$

"and so on." To make this idea more precise we introduce

DEFINITION 5. Let G be a group with neutral element e. For each $x \in G$, set

$$x^0 = e,$$

and for each nonnegative integer k, set

$$x^{k+1} = x^k \cdot x.$$

Then x^n is defined for all nonnegative integers, n, and for all $x \in G$; x^n is the n-*th power* of the element x.

For example,

if $k = 0$, then $x^1 = x^{0+1} = x^0 \cdot x = e \cdot x = x$;

if $k = 1$, then $x^2 = x^{1+1} = x^1 \cdot x = x \cdot x$;

if $k = 2$, then $x^3 = x^{2+1} = x^2 \cdot x = (x \cdot x) \cdot x$;

if $k = 3$, then $x^4 = x^{3+1} = x^3 \cdot x = ((x \cdot x) \cdot x) \cdot x$;

thus Definition 5 is in agreement with the practice of elementary algebra.

If the group operation is denoted by "$*$," the defining equations in Definition 5 become

$$x^0 = e,$$

and for each nonnegative integer k,

$$x^{k+1} = x^k * x.$$

In case the group G is written additively, i.e., with the group operation denoted by "$+$," Definition 5 is stated in the form

DEFINITION 5'. Let G (with group operation $+$) be a group with neutral element 0. For each $x \in G$ set

(3) $$0x = 0,$$

and for each nonnegative integer k, set

(4) $$(k + 1)x = kx + x.$$

Then nx is defined for all nonnegative integers n and for all $x \in G$. For each nonnegative integer n and for each $x \in G$, nx is the n-*th multiple* of x.

A few words of warning! Equations (3) and (4) must be regarded with a critical eye. The symbol "0" appearing on the left side of (3) stands for the *integer* zero; the symbol "0" appearing on the right side of (3) represents the neutral element of the group. In equation (4) the symbol "+" also does a double duty. The "+" appearing on the left side of (4) denotes addition of integers; the "+" appearing on the right side of (4) is the group operation in G.

Finally, the symbol "nx" (with $n \in Z$ and $x \in G$) does not mean "n times x." If n is positive, then "nx" stands for the sum

$$\underbrace{x + x + \cdots + x}_{n \text{ times}};$$

if n is zero, then $0x$ is defined by equation (3).

From the defining equations (3) and (4) of Definition 5′ we obtain: for $k = 0$,

$$1x = (0 + 1)x = 0x + x = 0 + x = x;$$

for $k = 1$,

$$2x = (1 + 1)x = 1x + x = x + x;$$

for $k = 2$,

$$3x = (2 + 1)x = 2x + x = (x + x) + x;$$

for $k = 3$,

$$4x = (3 + 1)x = 3x + x = ((x + x) + x) + x.$$

Throughout this text we shall denote the set of nonnegative integers by "N" or by "N_0"; thus

$$N = N_0 = \{n \mid n \in Z \text{ and } n \geq 0\}.$$

And if k is an integer, we set

$$N_k = \{n \mid n \in Z \text{ and } n \geq k\}.$$

The next two theorems[2] show that in groups one may operate with nonnegative integral exponents exactly as in elementary algebra.

THEOREM 4. If m and n are nonnegative integers, then for each $x \in G$,

$$x^{m+n} = x^m x^n.$$

---

[2] At a first reading of this text, it is a good idea to skip the proofs of Theorems 4, 5 and 6 as well as Exercises 2, 3 and 4, pages 65–67. These theorems are both familiar and plausible. Rather than getting bogged down in a mass of detail, it is preferable to get on with the business of group theory.

Proof: By the PFI. Let $x \in G$ and let

$$K = \{k \mid k \in N \text{ and } x^{m+k} = x^m x^k \text{ for each } m \in N\}.$$

We prove that $0 \in K$ and K is hereditary. This implies that K is the set N, whence the theorem follows.

Zero is in K since $x^{m+0} = x^m = x^m \cdot e = x^m \cdot x^0$. To prove that K is hereditary, we show: if $k \in K$, then $k + 1 \in K$. Thus, let $k \in K$; then

$$
\begin{aligned}
x^{m+(k+1)} &= x^{(m+k)+1} & &\text{(associativity of } + \text{ in } Z) \\
&= x^{m+k} \cdot x & &\text{(Definition 5)} \\
&= (x^m x^k)x & &\text{(since } k \in K) \\
&= x^m(x^k x) & &\text{(associativity of } \cdot) \\
&= x^m x^{k+1} & &\text{(Definition 5),}
\end{aligned}
$$

whence $k + 1 \in K$, and so K is hereditary. This completes the proof.

Q.E.D.

COROLLARY. If m and n are in N, then for each $x \in G$, $x^m x^n = x^n x^m$.

Proof: Exercise.

THEOREM 5. If m and n are in N, then for each $x \in G$,

$$(x^m)^n = x^{mn}.$$

Proof: Again we proceed by induction. Let

$$K = \{k \mid k \in N \text{ and } (x^m)^k = x^{mk} \text{ for each } m \in N\}.$$

First, $0 \in K$; for,

$$(x^m)^0 = e = x^{m \cdot 0}.$$

Now suppose $k \in K$; then

$$
\begin{aligned}
(x^m)^{k+1} &= (x^m)^k \cdot x^m & &\text{(Definition 5)} \\
&= x^{mk} \cdot x^m & &\text{(since } k \in K) \\
&= x^{mk+m} & &\text{(Theorem 4)} \\
&= x^{m(k+1)};
\end{aligned}
$$

the last step follows from the fact that multiplication is distributive over addition in Z. Therefore $k \in K$ implies $k + 1 \in K$, and so K is hereditary. By the PFI the theorem is proved.

Q.E.D.

Exercise. Restate Theorems 4 and 5 and the Corollary to Theorem 4 for additive groups and write out the proofs using additive notation.

In the foregoing we have defined, and discussed in some detail, non-negative powers of group elements. Next we wish to define negative powers of group elements. What shall we mean by the symbol "x^{-n}", where $n \in N$? Two reasonable definitions spring to mind: $x^{-n} = (x^{-1})^n$ and $x^{-n} = (x^n)^{-1}$. The following theorem resolves the dilemma of which to choose.

THEOREM 6.   For each element x in a group G and for each $n \in N$, $(x^n)^{-1} = (x^{-1})^n$.

> *Proof:* By definition, $(x^n)^{-1}$ is the inverse of x^n. If we can prove that $(x^{-1})^n$ is also an inverse of x^n, then since x^n has exactly one inverse it must be true that $(x^{-1})^n = (x^n)^{-1}$. Thus we need only show that for each $x \in G$ and for each nonnegative integer n
>
> $$(x^{-1})^n \cdot x^n = e.$$
>
> The proof proceeds by induction. Let
>
> $$K = \{k \mid k \in N \text{ and } (x^{-1})^k \cdot x^k = e\}.$$
>
> Clearly $0 \in K$. Now let $k \in K$; we show that $k + 1 \in K$. We have
>
> $$\begin{aligned}(x^{-1})^{k+1} \cdot x^{k+1} &= ((x^{-1})^k \cdot x^{-1}) \cdot (x \cdot x^k) \\ &= (x^{-1})^k \cdot (x^{-1} \cdot x) \cdot x^k \\ &= (x^{-1})^k \cdot e \cdot x^k \\ &= (x^{-1})^k \cdot x^k \\ &= e,\end{aligned}$$
>
> since $k \in K$. Hence K is hereditary, and so $K = N$.
>
> Q.E.D.

We are now in a position to define negative integral powers of elements in a group.

DEFINITION 6.   For each element x in a group G, and for each positive integer n,

$$x^{-n} = (x^{-1})^n;$$

x^{-n} is the $(-n)$-*th power* of the element x.

It follows from Theorem 6 and Definition 6 that

$$x^{-n} = (x^n)^{-1},$$

for each $x \in G$ and for each positive integer n.

Exercises

1. State Theorem 6 for additive groups and write out the proof using the additive notation.

2. State the definition corresponding to Definition 6 for additive groups.

3. Prove: If $x \in G$, then $x^n \in G$ for each integer n. (*Hint:* Use the PFI.)

4. Prove: For each element x in a group G and for all integers m and n,

(i) $$x^{-n} = (x^{-1})^n;$$

(ii) $$x^{m+n} = x^m \cdot x^n;$$

(iii) $$x^m \cdot x^n = x^n \cdot x^m;$$

(iv) $$(x^m)^n = x^{mn} = (x^n)^m.$$

5. Let x, y be elements in a group G such that $xy = yx$. Use the PFI to prove that $(xy)^n = x^n y^n$ for each $n \in Z$. (*Note:* We need not assume that G is Abelian.)

6. Use Exercise 5 to prove Theorem 6.

7. Carry out the proofs corresponding to Exercises 3, 4 and 5 for additive groups.

4. Order of a Group; Order of a Group Element

The few examples of groups discussed in the early part of this chapter indicate that we may separate groups into two classes: those which possess infinitely many elements and those which contain only a finite number of elements.

DEFINITION 7. A group G which contains only a finite number of elements is a *finite group*; the number of elements in a finite group G is the *order of G*. A group which contains infinitely many elements is an *infinite group*; such a group is said to have "infinite order."

Henceforth, if S is a finite set we shall denote the number of elements in S by "$|S|$." In particular, if G is a finite group, the order of G is written "$|G|$."

Next let G be a group and let x be any element in G. Consider the set of nonnegative powers of x,

(5) $$\{x^m \mid m \in N\}.$$

There are two mutually exclusive alternatives. On the one hand it may happen that for each pair of nonnegative integers n and m, if $n \neq m$, then $x^n \neq x^m$. In this case the set (5) contains infinitely many elements. Alternatively, there may exist a pair of integers n, m such that $n \neq m$ and

(6) $$x^m = x^n.$$

(Note that if G is a finite group, then since $x \in G$ there must be integers n and m, $n \neq m$, such that (6) holds. See Exercise 1, page 69 below.) Without loss of generality we may assume that $m > n$. Then from equation (6) it follows that

$$x^{m-n} = x^m \cdot x^{-n} = x^n \cdot x^{-n} = e$$

where $m - n$ is a positive integer. Thus, if there exist distinct nonnegative integers m, n such that $x^m = x^n$, then there is a positive integer k such that $x^k = e$. Let

$$P = \{k \mid k \in N \text{ and } x^k = e\};$$

by the Well-Ordering Principle, P contains a smallest positive integer r. Hence, from the existence of distinct nonnegative integers m and n such that $x^m = x^n$, we deduce that there is a smallest positive integer r such that $x^r = e$.

DEFINITION 8. An element $x \in G$ has *finite order* if and only if there is a positive integer n such that $x^n = e$. The smallest such integer is the *order* of x and is denoted by "$o(x)$." An element x in G has *infinite order* if and only if it does not have finite order.

Exercise. Write out Definition 8 for additive groups.

EXAMPLE 8. In Example 1 every element different from the neutral element has infinite order. The same is true of the elements different from the neutral element in Example 2. In Examples 3 and 4 it is easy to verify that every element has finite order. The neutral element of a group always has order 1.

May a group of infinite order have nonneutral elements of finite order? The next example shows that the answer is "Yes!"

EXAMPLE 9. Let $H = \{e, a\}$ be a two-element set, $a \neq e$, and let $\cdot$ be the binary operation on H defined by the table

TABLE 3

$\cdot$	e	a
e	e	a
a	a	e

Clearly, H together with $\cdot$ constitute a group. Now if Z is the additive group of integers, set

$$G = Z \times H;$$

thus $G = \{(n, x) \mid n \in Z \text{ and } x \in H\}$ and has infinitely many elements. Define $*$ on G by

$$(m, x) * (n, y) = (m + n, xy),$$

where $+$ is the operation of addition of integers and $\cdot$ (juxtaposition) is the operation on H defined by the table. The reader should verify that

(i) $*$ is a binary operation on G;

(ii) $(0, e)$ is a left neutral element; and

(iii) for each $(m, x) \in G$ the element $(-m, x^{-1})$ is a left inverse. (*Note:* for each $x \in H$, $x^{-1} = x$.)

Thus the system consisting of G and $*$ is a group. It is easy to check that $(0, e)$ and $(0, a)$ are both elements of finite order. In fact, these are the *only* elements in G of finite order. Later (Example 24, page 115) we shall give an example of an infinite group in which *every* element has finite order.

Exercises

1. Prove: If G is a finite group, then every element in G has finite order.

2. Let $Q_0 = Q - \{0\}$ where Q is the set of rational numbers. Verify: a) Q_0 is a group with respect to multiplication; b) the only elements of finite order in Q_0 are 1 and -1. What is $o(1)$? $o(-1)$?

We now use a few results from elementary number theory to gain some useful information concerning the orders of group elements. In Theorems 7 through 10 below, x is an element of a (multiplicative) group G.

THEOREM 7. Let m and n be integers where n is positive. If $o(x) = n$ and $x^m = e$, then $n \mid m$.

Proof: By the Division Algorithm, there is a unique pair of integers q and r such that
$$m = qn + r,$$
where $0 \le r < n$. By Exercise 4, page 67,
$$x^m = x^{qn+r} = (x^n)^q \cdot x^r.$$
But since $o(x) = n$, $x^n = e$, hence $(x^n)^q = e$, and so
$$x^m = e \cdot x^r = x^r;$$
by hypothesis we now have
$$x^r = e.$$
However, since n is the order of x, n is the smallest positive integer such that $x^n = e$. Since $0 \le r < n$ and since $x^r = e$, it follows that $r = 0$. Therefore $m = q \cdot n$, whence $n \mid m$.

Q.E.D.

THEOREM 8. Let n be a positive integer. Then $o(x) = n$ if and only if $o(x^{-1}) = n$.

Proof: Note, first, that if $x^k = e$ where k is a positive integer, then

$(x^{-1})^k = e$. For, $(x^{-1})^k$ is the inverse of x^k, hence

$$(x^{-1})^k = (x^{-1})^k \cdot e = (x^{-1})^k \cdot x^k = e.$$

The same argument shows that if $(x^{-1})^k = e$, then $x^k = e$.

Now let $o(x) = n$ and $o(x^{-1}) = m$. Then $x^n = e$, hence by the above argument $(x^{-1})^n = e$. By Theorem 7, $m \mid n$. Similarly, $(x^{-1})^m = e$ implies $x^m = e$ and therefore, by Theorem 7, $n \mid m$. Since $n \mid m$ and $m \mid n$, and both m and n are positive, we deduce $m = n$.

Q.E.D.

COROLLARY .   An element $x \in G$ has infinite order if and only if x^{-1} has infinite order.

Proof: Exercise.

THEOREM 9. Let m and n be integers, where $n > 0$, and let $d = (m, n)$ be the greatest common divisor of m and n, and let $o(x) = n$. Then $o(x^m) = n/d$.

Proof: Since $(m, n) = d$, $m = r \cdot d$ and $n = s \cdot d$ where r and s are integers such that $(s, r) = 1$.

Now

$$(x^m)^{n/d} = (x^{r \cdot d})^{n/d} = (x^n)^r = e^r = e,$$

so that x^m has finite order. If $q = o(x^m)$, Theorem 7 shows that $q \mid n/d$. On the other hand, $o(x^m) = q$ implies

$$(x^m)^q = x^{mq} = e.$$

Hence, by Theorem 7, $n \mid mq$, i.e., $sd \mid rdq$ or $s \mid rq$. But $(s, r) = 1$; therefore $s \mid q$, i.e., $n/d \mid q$. Thus n/d and q are positive integers such that $n/d \mid q$ and $q \mid n/d$. Therefore $o(x^m) = q = n/d$.

Q.E.D.

COROLLARY. Let m and n be integers, $n > 0$, and let $o(x) = n$. Then $o(x^m) = n$ if and only if $(m, n) = 1$.

Proof: Exercise.

THEOREM 10.   Let n be a positive integer. Then $o(x) = n$ if and only if
  (i) the elements $x^0 (= e), x^1, \ldots, x^{n-1}$ are distinct, and
  (ii) $\{x^m \mid m \in Z\} = \{x^0, x^1, \ldots, x^{n-1}\}$.

Proof: Let $o(x) = n$; we prove that (i) and (ii) hold.
        (i) Suppose $x^i = x^j$ where $0 \leq i \leq n - 1$, $0 \leq j \leq n - 1$.

We may assume that $j < i$; then setting $k = i - j$, we have $0 < k \leq n - 1$, and

$$x^k = x^{i-j} = e.$$

But n is the smallest positive integer such that $x^n = e$. Thus we have a contradiction, and therefore the assumption is false. This proves (i).

(ii) To prove (ii), note that $\{x^0, x^1, \ldots, x^{n-1}\} \subset \{x^m \mid m \in Z\}$. Therefore we need only demonstrate that $\{x^m \mid m \in Z\} \subset \{x^0, x^1, \ldots, x^{n-1}\}$. Let $x^k \in \{x^m \mid m \in Z\}$. By the Division Algorithm there exists a unique pair of integers q and r such that $k = qn + r$ where $0 \leq r < n$. Then, since $o(x) = n$, by the argument used in Theorem 7,

$$x^k = x^r;$$

but $0 \leq r < n$ implies $x^r \in \{x^0, x^1, \ldots, x^{n-1}\}$. Hence $x^k \in \{x^0, x^1, \ldots, x^{n-1}\}$ and (ii) is proved.

Conversely, suppose (i) and (ii) hold. By (i) it follows that $o(x) \geq n$. But, by (ii), $x^n \in \{x^0, x^1, \ldots, x^{n-1}\}$. If $x^n = x^i$, $0 \leq i \leq n - 1$, then $x^{n-i} = e$, whence $o(x) \leq n - i \leq n$. Since we have already seen that $o(x) \geq n$, we conclude $o(x) = n$.

Q.E.D.

COROLLARY. If n is a positive integer and $o(x) = n$, then the number of elements in the set $\{x^m \mid m \in Z\}$ is exactly n.

Proof: Trivial.

Exercises

1. Let z be an element in a group G and let $o(z) = mn$. Prove that there exist elements $a, b \in G$ such that $ab = ba$ and $o(a) = n$, $o(b) = m$. *Hint:* Let $a = z^m$ and $b = z^n$.

2. Let x, y be elements in a group G of orders m and n, respectively, such that $xy = yx$. Prove that the order of xy is mn if and only if $(m, n) = 1$. *Hint:* Use Exercise 5, page 67.

3. State and write out the proofs of Theorems 7 through 10 for additive groups.

5. Cyclic Groups

The simplest groups are those in which every group element "can be obtained" from some one of them. Such groups are termed "cyclic groups." To illustrate the concept we present:

EXAMPLE 10. Consider the roots of the equation $x^4 - 1 = 0$. These roots comprise the set $G = \{1, -1, i, -i\}$ where $i^2 = -1$. With multiplication of complex numbers as the binary operation, G is a group with neutral element 1. Note that

$$i^0 = 1, \qquad i^1 = i, \qquad i^2 = -1, \qquad i^3 = -i;$$

in fact, every integral power of i is some one of the elements $1, -1, i, -i$. Thus G consists of all the integral powers of one of its elements, i. G is a *cyclic* group *generated* by the element i. Clearly G is also generated by the element $-i$. This group is the *group of the fourth roots of unity.*

DEFINITION 9. A (multiplicative) group G is *cyclic* if and only if there is an $x \in G$ such that each $y \in G$ is an integral power of x. The element x is a *generator* of G and we write

$$G = (x).$$

Alternatively, Definition 9 may be stated thus: A (multiplicative) group G is a *cyclic* group with *generator* $x \in G$ if and only if the set of elements comprising G is the set

$$\{x^m \mid m \in Z\}.$$

DEFINITION 9'. A (additive) group is *cyclic* if and only if there is an $x \in G$ such that each $y \in G$ is an integral multiple of x. As above, x is a *generator* of G and we write "$G = (x)$."

Definition 9' may also be stated: A (additive) group G is a *cyclic* group with *generator* $x \in G$ if and only if the set of elements comprising G is the set

$$\{mx \mid m \in Z\}.$$

EXAMPLE 11. The additive group of integers Z is a cyclic group with 1 as a generator. For, each $y \in Z$ is a multiple of 1; i.e., if $y \in Z$, then

$$y = y \cdot 1.$$

Examples 9 and 10 show that a cyclic group may be finite or infinite. Our next theorem characterizes finite (and therefore, also, infinite) cyclic groups by properties of powers of a generator.

THEOREM 11. Let G be a cyclic group with generator x, $G = (x)$.

(i) If x has finite order, so does G and $|G| = o(x)$.
(ii) If G has finite order, so does x and again $|G| = o(x)$.

Proof: (i) Let $o(x) = n$ be a positive integer. By Theorem 10, $x^0, x^1,$ $\ldots, x^{n-1}$ are distinct and $\{x^0, x^1, \ldots, x^{n-1}\} = \{x^m \mid m \in Z\}$. But since G is a cyclic group with generator x, $G = \{x^m \mid m \in Z\}$. Hence $G = \{x^0, x^1, \ldots, x^{n-1}\}$ and so $|G| = n$.

(ii) Suppose G has finite order. Since $G = \{x^m \mid m \in Z\}$ there exist distinct integers i and j such that $x^i = x^j$. Without loss of generality we may assume $i > j$. Then

$$x^{i-j} = x^i \cdot x^{-j} = x^j \cdot x^{-j} = e,$$

i.e., some positive power of x is the neutral element. By Definition 8, x has finite order and by part (i) it now follows that $|G| = o(x)$.

Q.E.D.

COROLLARY 1. A cyclic group G with generator x is finite if and only if there exist distinct integers i and j such that $x^i = x^j$.

Proof: Exercise.

COROLLARY 2. A cyclic group G with generator x has infinite order if and only if x has infinite order.

Proof: Exercise.

Example 10 shows that a cyclic group may have more than one generator. The next theorem provides a criterion for an element in a cyclic group to be a generator of the group.

THEOREM 12. Let $G = (x)$ and let $y \in G$ where $y = x^m$.

(i) If $|G| = n$, then y is a generator of G if and only if n and m are relatively prime.

(ii) If G has infinite order, then y is a generator of G if and only if $y = x$ or $y = x^{-1}$.

Proof: (i) The element y is a generator of G if and only if $o(y) = n$, hence if and only if $o(x^m) = n$. By the Corollary to Theorem 9, $G = (y)$ if and only if m and n are relatively prime.

(ii) Let $G = (x)$ where G has infinite order. Clearly $G = (x^{-1})$ (why?). We claim that G has no generator other than x and x^{-1}.

For suppose $G = (y)$ where $y = x^m$ and m is an integer. If $m = 1$, then $y = x$, and if $m = -1$, then $y = x^{-1}$. It now suffices to show that if $m \neq 1$ and $m \neq -1$, we are led to a contradiction. Since $x \in G$ and $G = (y)$, $x = y^t$ for some integer t. Then

$$x = y^t = (x^m)^t = x^{mt}.$$

Since $m \neq 1$ and $m \neq -1$ and since t is an integer, $mt \neq 1$. Thus two distinct powers of x, namely, $x = x^1$ and x^{mt}, are equal. By Corollary 1 to Theorem 11, G must be finite, a contradiction.

Q.E.D.

We have seen that the group of the fourth roots of unity (Example 10) is cyclic with i as generator. Since $o(i) = 4$ and since $-i = i^3$, and since

$(4, 3) = 1$, it follows from Theorem 12 that $-i$ is also a generator of the group. On the other hand, $-1 = i^2$ is not a generator since $(4, 2) = 2$.

The additive group of integers is, of course, an infinite cyclic group having only 1 and -1 as generators.

Exercises

1. Prove that every cyclic group is Abelian.

2. Let G be an infinite cyclic group. Prove that there is a one-one correspondence between G and the set Z of all integers.

3. Prove that the cube roots of unity (i.e., the roots of $x^3 - 1 = 0$) form a cyclic group of order three.

4. Let G be a finite cyclic group of order n, say, $G = (x)$. Prove that $x^{-1} = x^{n-1}$.

5. Rewrite Theorems 11 and 12 and their proofs for additive groups.

6. Let $|G| = n$ where n is a positive integer. Prove that G is cyclic if and only if there is an $x \in G$ such that $o(x) = n$.

7. Prove that if $(ab)^2 = a^2b^2$ for each a and b in G, then G is Abelian.

8. Let G be a group of order n, and let p be a prime dividing n. Prove that there is an element $x \in G$ such that $o(x) = p$.

6. The Symmetric Groups

The *symmetric groups*, and their generalizations, the *full transformation groups*, are rich sources of examples of groups of all kinds. The full truth of this statement will become apparent from Theorem 4, Chapter 3. The present section is restricted to a study of the symmetric groups; some of the considerations presented here will be extended, in Section 8, to the full transformation groups.

DEFINITION 10. Let $A_n = \{x_1, x_2, \ldots, x_n\}$, $n \geq 1$, be a set with n elements. A *permutation* of A_n is a bijection $A_n \longrightarrow A_n$. The set of all permutations of A_n is denoted by "$\mathscr{S}(A_n)$" or, more briefly, by "$\mathscr{S}_n$."

THEOREM 13. The set $\mathscr{S}_n$ is a group with respect to composition of mappings as group operation.

Proof: If φ and ψ are in $\mathscr{S}_n$, then φ and ψ are bijections of A_n. Hence so is $\varphi \circ \psi$ and therefore $\varphi \circ \psi \in \mathscr{S}_n$. Thus $\circ$ is a binary operation on $\mathscr{S}_n$.

Next, $\circ$ is associative (recall: composition of mappings is always associative.) Further, the mapping ι defined by $\iota(x) = x$ for each $x \in A_n$ is the neutral element of $\mathscr{S}_n$. Finally, if $\varphi \in \mathscr{S}_n$, then $\varphi^{-1} \in \mathscr{S}_n$ and $\varphi^{-1} \circ \varphi = \iota$. Hence the structure consisting of $\mathscr{S}_n$ and $\circ$ is a group.

Q.E.D.

It can be shown that $\mathscr{S}_n$ is a finite group; indeed, in Exercise 2, page 82, you are asked to prove that $|\mathscr{S}_n| = n!$.

DEFINITION 11. $\mathscr{S}_n$ is the *symmetric group on n elements*.

To facilitate our study of the symmetric groups, it will be convenient to adopt the standard notations and terminology.
1. If $\varphi, \psi \in \mathscr{S}_n$, then henceforth we write "$\varphi\psi$" instead of "$\varphi \circ \psi$"; we shall speak of the "product of φ and ψ" rather than the "composite of φ and ψ."
2. Suppose $\varphi \in \mathscr{S}_n$; a useful notation for the representation of φ is

$$(7) \qquad \begin{pmatrix} x_1 & x_2 & \cdots & x_n \\ \varphi(x_1) & \varphi(x_2) & \cdots & \varphi(x_n) \end{pmatrix}.$$

For example, if $\varphi \in \mathscr{S}_5$ is defined by

$$\varphi(x_1) = x_3,\ \varphi(x_2) = x_1,\ \varphi(x_3) = x_2,\ \varphi(x_4) = x_4,\ \varphi(x_5) = x_5,$$

then

$$\varphi = \begin{pmatrix} x_1 & x_2 & x_3 & x_4 & x_5 \\ x_3 & x_1 & x_2 & x_4 & x_5 \end{pmatrix};$$

each symbol in the second row represents the image, under φ, of the element represented directly above it in the first row. From this description it follows that

$$\begin{pmatrix} x_1 & x_2 & x_3 & x_4 & x_5 \\ x_3 & x_1 & x_2 & x_4 & x_5 \end{pmatrix} = \begin{pmatrix} x_2 & x_1 & x_3 & x_5 & x_4 \\ x_1 & x_3 & x_2 & x_5 & x_4 \end{pmatrix} = \text{etc.}$$

In general, if $\varphi \in \mathscr{S}_n$ is given by (7), then

$$\begin{pmatrix} x_1 & x_2 & \cdots & x_n \\ \varphi(x_1) & \varphi(x_2) & \cdots & \varphi(x_n) \end{pmatrix} = \begin{pmatrix} x_2 & x_1 & \cdots & x_{n-1} & x_n \\ \varphi(x_2) & \varphi(x_1) & \cdots & \varphi(x_{n-1}) & \varphi(x_n) \end{pmatrix} = \text{etc.}$$

Our list of notations is not yet complete. In a little while we shall define a special kind of permutation, a "cycle," and the notation for cycles is particularly simple. Moreover, it will turn out that cycles play an essential role in the theory of the symmetric groups.

In order to investigate the groups $\mathscr{S}_n$, we wish to examine the effects of the elements of $\mathscr{S}_n$ upon the elements of A_n.

DEFINITION 12.   Let $\varphi \in \mathscr{S}_n$ and let $x, y \in A_n$. Then x is *φ-equivalent* to y if and only if there is an integer h such that $\varphi^h(x) = y$. We write

$$x \sim_\varphi y.$$

Clearly, $\sim_\varphi$ is an equivalence relation on A_n (details?), hence $\sim_\varphi$ partitions A_n.

DEFINITION 13. The equivalence classes of the partition determined by φ are the *orbits* of φ. An orbit of φ is *trivial* if and only if it consists of one element of A_n; otherwise it is *nontrivial*.

The only permutation having trivial orbits exclusively is the identity permutation, ι.

EXAMPLE 12.   Let $\varphi \in \mathscr{S}_{10}$ be the permutation

$$\varphi = \begin{pmatrix} x_1 & x_2 & x_3 & x_4 & x_5 & x_6 & x_7 & x_8 & x_9 & x_{10} \\ x_{10} & x_5 & x_6 & x_4 & x_9 & x_7 & x_8 & x_3 & x_2 & x_1 \end{pmatrix}.$$

The orbits of φ are

$$\{x_4\}, \{x_1, x_{10}\}, \{x_2, x_5, x_9\}, \{x_3, x_6, x_7, x_8\},$$

and of these only the orbit $\{x_4\}$ is trivial.

Exercises

1. Determine the orbits of each of the following permutations:

a) $\begin{pmatrix} x_1 & x_2 & x_3 & x_4 & x_5 & x_6 & x_7 \\ x_3 & x_4 & x_2 & x_1 & x_7 & x_6 & x_5 \end{pmatrix}$;

b) $\begin{pmatrix} x_1 & x_2 & x_3 & x_4 & x_5 & x_6 & x_7 & x_8 & x_9 & x_{10} \\ x_9 & x_2 & x_4 & x_6 & x_7 & x_3 & x_8 & x_5 & x_1 & x_{10} \end{pmatrix}$;

c) the permutation $\begin{pmatrix} x & y \\ y & x \end{pmatrix}$.

2. Prove that $x \in A_n$ is an element in a nontrivial orbit of φ if and only if $\varphi(x) \neq x$.

3. Does the set of orbits of a permutation determine the permutation? Why? (Note that a permutation φ completely determines its orbits.)

4. Let $\varphi \in \mathscr{S}_n$ and let $x \in A_n$. Prove that the orbit of φ containing x is the set $\{\varphi^m(x) \mid m \in N\}$. Also, show that there is a smallest $r \in N$ such that $\varphi^r(x) = x$ and therefore the orbit of φ containing x is

$$\{x, \varphi(x), \ldots, \varphi^{r-1}(x)\}.$$

7. *Cycles; Decomposition of Permutations Into Disjoint Cycles*

The cycles are the simplest permutations, namely, those having fewest nontrivial orbits.

DEFINITION 14.   A permutation $\gamma \in \mathscr{S}_n$ is a *cycle* if and only if it has at most one nontrivial orbit.

Evidently the identity permutation, ι, is a cycle; in fact, it is the only cycle having no nontrivial orbit.

Since a cycle γ is an element of the finite group $\mathscr{S}_n$, it has finite order, say, $o(\gamma) = k$. In this case we say that γ is a "k-cycle." Two-cycles are called "transpositions" and the only 1-cycle is ι.

EXAMPLE 13. Let $\gamma \in \mathscr{S}_5$ be the permutation

$$\gamma = \begin{pmatrix} x_1 & x_2 & x_3 & x_4 & x_5 \\ x_3 & x_2 & x_4 & x_5 & x_1 \end{pmatrix}.$$

The orbits of γ are

$$\{x_2\} \qquad \text{and} \qquad \{x_1, x_3, x_4, x_5\},$$

and since it has one nontrivial orbit, γ is a cycle. The order of γ is four (why?), hence γ is a 4-cycle. Schematically, γ is described by the cyclic diagram of Figure 8.

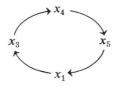

FIGURE 8

Tracing the diagram in the indicated direction from each symbol to the next, one finds the image of each of the elements x_1, x_3, x_4, x_5. The absence of x_2 from the diagram is interpreted to mean that $\varphi(x_2) = x_2$.

From the cyclic diagram of Figure 8, we derive a simple linear notation for cycles. If we break the cyclic diagram before the "x_1" and stretch it out, so

(8) $(x_1 \quad x_3 \quad x_4 \quad x_5)$,

then we interpret (8) as meaning that

$$\gamma(x_1) = x_3, \ \gamma(x_3) = x_4, \ \gamma(x_4) = x_5, \ \gamma(x_5) = x_1.$$

More generally, suppose $\gamma \in \mathscr{S}_n$ is a cycle having the nontrivial orbit $\mathcal{O}_\gamma = \{y_1, y_2, \ldots, y_k\}$ where $y_i \in A_n$, $1 \leq i \leq k$, and

$$y_2 = \gamma(y_1), \ y_3 = \gamma(y_2), \ldots, y_k = \gamma(y_{k-1}), \ y_1 = \gamma(y_k);$$

then γ is usually represented by

(9) $(y_1 \quad y_2 \quad \cdots \quad y_k)$.

The representation (9) is not uniquely determined by γ. For, in defining (9), we began with the element $y_1 \in \mathcal{O}_\gamma$. Since the choice of starting element is completely arbitrary, we might just as well begin with y_2. This choice leads to the representation of γ by

$$(y_2 \quad y_3 \quad \cdots \quad y_k \quad y_1).$$

In general, we have

$$
\begin{aligned}
(y_1 \quad y_2 \quad \cdots \quad y_k) &= (y_2 \quad y_3 \quad \cdots \quad y_k \quad y_1) \\
&= (y_3 \quad y_4 \quad \cdots \quad y_k \quad y_1 \quad y_2) \\
&= \text{etc.}
\end{aligned}
$$

On the other hand, $(y_1 \quad y_2 \quad y_3 \quad \cdots \quad y_k) \neq (y_2 \quad y_1 \quad y_3 \quad \cdots \quad y_k)$, etc.

The identity permutation is usually denoted by "(x_1)," although "$(\ )$" might be a better notation.

Example 13, and other examples of cycles that the reader may have examined, suggests that the number of elements in the nontrivial orbit of a cycle $\neq \iota$ is equal to the order of the cycle. This conjecture is now verified in

THEOREM 14. Let $\gamma \neq \iota$ be a cycle in $\mathscr{S}_n$, and let $\mathcal{O}_\gamma$ be its nontrivial orbit. Further, let $|\mathcal{O}_\gamma|$ be the number of elements in the orbit $\mathcal{O}_\gamma$. Then $|\mathcal{O}_\gamma| = o(\gamma)$.

Proof: Let $\mathcal{O}_\gamma = \{y_1, y_2, \ldots, y_k\}$, $k > 1$, where the y's are distinct and where

$$y_2 = \gamma(y_1), \ y_3 = \gamma(y_2), \ldots, y_k = \gamma(y_{k-1}), \ y_1 = \gamma(y_k).$$

Then

$$\gamma^k(y_1) = \gamma^{k-1}(y_2) = \cdots = \gamma(y_k) = y_1,$$

and similarly

$$\gamma^k(y_i) = y_i, \quad 1 \leq i \leq k.$$

Moreover, if x is an element in a trivial orbit of γ, then $\gamma(x) = x$, therefore $\gamma^k(x) = x$. Hence

$$\gamma^k(z) = z,$$

for each $z \in A_n$ and so $\gamma^k = \iota$. Consequently (Theorem 7), $o(\gamma) \mid k$.

Let $o(\gamma) = j$; then $1 < j \leq k$. We prove that $j < k$ is impossible. Indeed, if $j < k$, then

$$y_2 = \gamma(y_1), \ y_3 = \gamma(y_2) = \gamma^2(y_1), \ldots, y_j = \gamma^{j-1}(y_1)$$

and

$$y_{j+1} = \gamma^j(y_1) = y_1.$$

But $1 < j < k$ implies that y_{j+1} is one of the elements in the set $\{y_1, y_2, \ldots, y_k\}$, which were, in the first place, assumed distinct. Thus the hypothesis $o(\gamma) = j < k$ yields a contradiction and therefore $o(\gamma) = k$.

Q.E.D.

COROLLARY. If $\gamma \in \mathscr{S}_n$ is a k-cycle, then for each element y in the nontrivial orbit $\mathcal{O}_\gamma$ of γ,

$$\mathcal{O}_\gamma = \{\gamma^j(y) \mid j = 0, 1, \ldots, k - 1\}.$$

Proof: Exercise.

Now let $\varphi \in \mathscr{S}_n$ and let $x \in A_n$. By Exercise 2, page 76, x is in a nontrivial orbit of φ if and only if $\varphi(x) \neq x$. This fact suggests

DEFINITION 15. The *displaced* set D_φ of φ is the union of the nontrivial orbits of φ.

In view of the remark preceding Definition 15,

$$D_\varphi = \{x \mid x \in A_n \text{ and } \varphi(x) \neq x\}.$$

If φ is a cycle, D_φ is the nontrivial orbit of φ.

DEFINITION 16. Two permutations φ, ψ in $\mathscr{S}_n$ are *disjoint* if and only if $D_\varphi \cap D_\psi = \emptyset$.

Exercises

In the exercises below φ, ψ, and $\psi_1, \psi_2, \ldots, \psi_m$ are elements in $\mathscr{S}_n$. Prove:

1. $\mathscr{S}_n$ has finite order, hence φ has finite order.
2. The inverse of a k-cycle is a k-cycle.
3. If $n < 4$, then each permutation in $\mathscr{S}_n$ is a cycle.
4. If φ and ψ are disjoint, then $D_{\varphi\psi} = D_\varphi \cup D_\psi$.
5. φ and ψ are disjoint if and only if for each $y \in D_\varphi$, $\{y\}$ is a trivial orbit of ψ and for each $z \in D_\psi$, $\{z\}$ is a trivial orbit of φ.
6. If $y \in D_\varphi$, then $\varphi(y) \in D_\varphi$.
7. φ is a transposition if and only if φ has exactly $n - 1$ orbits. (In this case, φ has exactly one nontrivial orbit and this orbit is a two-element set.)
8. If $\varphi = \psi_1\psi_2 \cdots \psi_m$ where for $i \neq j$, ψ_i and ψ_j are disjoint, then $D_\varphi = D_{\psi_1} \cup D_{\psi_2} \cup \cdots \cup D_{\psi_m}$. (*Note:* In case the ψ's are all cycles, the D_ψ's are all the nontrivial orbits of φ.)
9. If φ and ψ are disjoint and both different from ι, then $\varphi\psi$ is not a cycle.

In general, if φ and ψ are arbitrary permutations, then $\varphi\psi \neq \psi\varphi$. For example,

$$(x_1x_2)(x_1x_3) = (x_1x_3x_2) \neq (x_1x_2x_3) = (x_1x_3)(x_1x_2).$$

However, if the permutations are disjoint we have

THEOREM 15. If φ and ψ are disjoint permutations in $\mathscr{S}_n$, then $\varphi\psi = \psi\varphi$.

Proof: We wish to verify that for each $y \in A_n$

$$(\varphi\psi)(y) = (\psi\varphi)(y).$$

CASE 1. $y \in D_\varphi$. Then (Exercise 5, above) $\{y\}$ is a trivial orbit of ψ and so $\psi(y) = y$. Hence

$$(\varphi\psi)(y) = \varphi(y).$$

But $y \in D_\varphi$ imples $\varphi(y) \in D_\varphi$ (Exercise 6, page 79). Therefore $\{\varphi(y)\}$ is a trivial orbit of ψ and consequently

$$(\psi\varphi)(y) = \psi(\varphi(y)) = \varphi(y).$$

Therefore

$$(\varphi\psi)(y) = (\psi\varphi)(y).$$

CASE 2. $y \in D_\psi$. Reasoning as in Case 1 yields the same result.

CASE 3. $y \notin D_\varphi \cup D_\psi$. Then $\{y\}$ is a trivial orbit of both φ and ψ, whence

$$(\varphi\psi)(y) = y = (\psi\varphi)(y).$$

Q.E.D.

The main theorem of this section shows that the cycles in $\mathscr{S}_n$ are building blocks of $\mathscr{S}_n$.

THEOREM 16.   Every permutation $\varphi \in \mathscr{S}_n$ is a product of pairwise disjoint cycles. In other words, there exist cycles $\gamma_1, \gamma_2, \ldots, \gamma_m$, $m \geq 1$, in $\mathscr{S}_n$ such that each pair $\gamma_i, \gamma_j, i \neq j$, is disjoint and such that

$$\varphi = \gamma_1\gamma_2 \cdots \gamma_m.$$

Moreover, $\gamma_1, \gamma_2, \ldots, \gamma_m$ are uniquely determined up to order. That is to say: If

$$\gamma_1\gamma_2 \cdots \gamma_m = \eta_1\eta_2 \cdots \eta_r$$

where η's are also pairwise disjoint and where no γ and no η is the identity permutation, then

(i) $r = s$, and
(ii) each $\gamma_i = $ some η_j and each $\eta_j = $ some γ_i.

Remark.   In almost every specific case, the factorization of a permutation into disjoint cycles is evident by inspection. For instance, if $\varphi \in \mathscr{S}_{10}$ is the permutation

$$\varphi = \begin{pmatrix} x_1 & x_2 & x_3 & x_4 & x_5 & x_6 & x_7 & x_8 & x_9 & x_{10} \\ x_{10} & x_5 & x_6 & x_4 & x_9 & x_7 & x_8 & x_3 & x_2 & x_1 \end{pmatrix},$$

we see at once that

$$\varphi = (x_1 x_{10})(x_2 x_5 x_9)(x_3 x_6 x_7 x_8),$$

and that this is the only factorization of φ into disjoint cycles. However, when one attempts to give a formal description of the intuitive process used to factor φ, then the situation becomes quite messy. For those who will be satisfied with nothing less than a formal proof we begin with a:

LEMMA.   Let $\mathcal{O}$ be a nontrivial orbit of φ and let $x \in \mathcal{O}$. Define $\gamma : A_n \longrightarrow A_n$ by

$$\gamma(y) = \begin{cases} \varphi(y), \text{ if } y \in \mathcal{O} \\ y, \text{ if } y \notin \mathcal{O}. \end{cases}$$

Then γ is a cycle whose nontrivial orbit is $\mathcal{O}$.

Proof: It suffices to show that $\mathcal{O}$ is the only nontrivial orbit of γ (see Definition 14). Let $u, v \in \mathcal{O}$; since $\mathcal{O}$ is an orbit of φ, there is an integer h such that $\varphi^h(u) = v$. But $\gamma(u) = \varphi(u)$ and so $\gamma^h(u) = \varphi^h(u)$. Consequently $\gamma^h(u) = v$ and therefore $u \sim_\gamma v$. Thus the elements of $\mathcal{O}$ are all in one orbit of γ. On the other hand, if $w \notin \mathcal{O}$, then $\gamma(w) = w$ (definition of γ), whence $\mathcal{O}$ is the only nontrivial orbit of γ.

Q.E.D.

Proof of Theorem 16.   Let $\mathcal{O}_1, \mathcal{O}_2, \ldots, \mathcal{O}_m$ be the nontrivial orbits of φ and define $\gamma_i : A_n \longrightarrow A_n$ by

$$\gamma_i(y) = \begin{cases} \varphi(y), \text{ if } y \in \mathcal{O}_i \\ y, \text{ if } y \notin \mathcal{O}_i, \ 1 \le i \le m. \end{cases}$$

By the Lemma, each γ_i is a cycle with $\mathcal{O}_i$ as its nontrivial orbit. The displaced set of γ_i is $\mathcal{O}_i$ and for $i \ne j$, $\mathcal{O}_i \cap \mathcal{O}_j = \emptyset$. Hence (Theorem 15) the γ's commute in pairs. We claim that

$$(10) \qquad\qquad \varphi = \gamma_1 \gamma_2 \cdots \gamma_m,$$

and to verify this result it suffices to prove

$$(11) \qquad\qquad \varphi(y) = (\gamma_1 \gamma_2 \cdots \gamma_m)(y),$$

for each $y \in A_n$. Now if $y \notin D_\varphi$, then (Definition 15, page 79) $y \notin \mathcal{O}_i$ for each i, $1 \le i \le m$, and therefore $\varphi(y) = y$. But $y \notin \mathcal{O}_i$, for each i, also implies $\gamma_i(y) = y$, $1 \le i \le m$. Hence

$$\varphi(y) = (\gamma_1 \gamma_2 \cdots \gamma_m)(y)$$

in case $y \notin D_\varphi$. Suppose $y \in D_\varphi$; since D_φ is the union of the $\mathcal{O}$'s and since the $\mathcal{O}$'s are pairwise disjoint, there is exactly one $\mathcal{O}_j$ such that $y \in \mathcal{O}_j$. Then

$$\gamma_j(y) = \varphi(y),$$

and for each $i \ne j$,

$$\gamma_i(y) = y.$$

Hence

$$(\gamma_1 \cdots \gamma_j \cdots \gamma_m)(y) = (\gamma_j(\gamma_1 \cdots \gamma_{j-1}\gamma_{j+1} \cdots \gamma_n))(y)$$
$$= \gamma_j(y)$$
$$= \varphi(y).$$

Thus, in every case, equation (11) holds and so (10) is verified. Suppose

$$\varphi = \gamma_1\gamma_2 \cdots \gamma_m = \eta_1\eta_2 \cdots \eta_r,$$

where the η's are also pairwise disjoint cycles. Then the nontrivial orbits of each of the η_i is also a nontrivial orbit of φ (Exercise 8, page 79). Since φ completely determines its orbits, it follows that the orbits of the η's are $\mathcal{O}_1, \mathcal{O}_2, \ldots, \mathcal{O}_m$ and therefore $r = m$. Further, the nontrivial orbit of each η_i is the nontrivial orbit of some γ_j. After relabeling, we may assume that η_i and γ_i have the same nontrivial orbit, $1 \leq i \leq m$. We prove that

$$\eta_i = \gamma_i, \ 1 \leq i \leq m.$$

If $y \in A_n$ and $y \in$ some $\mathcal{O}_j$, then

$$\eta_k(y) = \gamma_k(y) = y$$

for $k \neq j$, and

$$\eta_j(y) = \varphi(y) = \gamma_j(y).$$

Hence, for each $y \in A_n$ and for each i, $1 \leq i \leq m$,

$$\eta_i(y) = \gamma_i(y),$$

whence $\eta_i = \gamma_i, \ 1 \leq i \leq m$.

Q.E.D.

Our study of the symmetric groups was based upon the selection, in advance, of a set $A_n = \{x_1, x_2, \ldots, x_n\}$, $n \geq 1$, of n elements. Suppose, now, some other n-element set $B_n = \{y_1, y_2, \ldots, y_n\}$ were chosen as the underlying set. It is clear that with B_n as base of operations a symmetric group $\mathscr{S}'_n$ can be defined and all the theorems of the present section carry over to $\mathscr{S}'_n$. Since the nature of the elements in A_n has nowhere entered our reasoning, it is plausible that $\mathscr{S}_n$ and $\mathscr{S}'_n$ are algebraically indistinguishable. In Chapter 3, the concept of algebraic indistinguishability or *isomorphism* will be made precise and it will be proved that $\mathscr{S}_n$ and $\mathscr{S}'_n$ are isomorphic.

Exercises

1. Let $\varphi = (a_1a_2 \cdots a_{m_1})(b_1b_2 \cdots b_{m_2}) \cdots (c_1c_2 \cdots c_{m_k})$ where $\varphi \in \mathscr{S}_n$ and the cycles are disjoint and nontrivial. Prove that $\{a_1, a_2, \ldots, a_{m_1}\}$, $\{b_1, b_2, \ldots, b_{m_2}\}$, $\ldots, \{c_1, c_2, \ldots, c_{m_k}\}$ are all the nontrivial orbits of φ. Also, prove that the number of orbits, $N(\varphi)$, is $N(\varphi) = k + r$ where $r = n - (m_1 + m_2 + \cdots + m_k)$.

2. Define $F_i = \{\varphi \mid \varphi \in \mathscr{S}_n$ and $\varphi(x_n) = x_i\}$, $1 \leq i \leq n$, where $n > 1$. Prove:
(a) There exists a bijection $F_n \longrightarrow \mathscr{S}_{n-1}$.
(b) For each i and each j, there is a bijection $F_i \longrightarrow F_j$.
(c) $\bigcup_{i=1}^{n} F_i = \mathscr{S}_n$ and if $i \neq j$, $F_i \cap F_j = \emptyset$. Hence deduce that $|\mathscr{S}_n| = n!$.

3. (a) Verify that $(x_ix_jx_k) = (x_ix_k)(x_ix_j)$.
(b) Show that every k-cycle $(x_{i_1}x_{i_2} \cdots x_{i_k})$ has a factorization

$$(x_{i_1}x_{i_2} \cdots x_{i_k}) = (x_{i_1}x_{i_k})(x_{i_1}x_{i_2} \cdots x_{i_{k-1}}).$$

(c) Use the PFI to prove that each permutation is a product of transpositions.

(d) Verify that $(x_i x_j) = (x_1 x_i)(x_1 x_j)(x_1 x_i)$; hence prove that every permutation in $\mathscr{S}_n$ is a product of transpositions in the set $\{(x_1 x_i) \mid 2 \leq i \leq n\}$.

8. Full Transformation Groups

The symmetric group $\mathscr{S}_n$ was defined as the set of all one-one mappings of a *finite* set $A_n = \{x_1, x_2, \ldots, x_n\}$ onto itself. One might ask: How much of the theory developed depended upon the restriction to *finite* underlying sets? It turns out that if this restriction is lifted (i.e., if the underlying sets are allowed to be either finite or infinite), many of the theorems of Sections 6 and 7 hold with little or even no modification. Since this is the case, the proofs of the theorems in this section will only be indicated, all details being left to the reader.

DEFINITION 17.   A *permutation on* A is a one-one mapping of the set A onto itself. The set of all permutations on A is denoted by "$\mathscr{P}(A)$."

THEOREM 17.   The set $\mathscr{P}(A)$ is a group with respect to composition of mappings as group operation.

Proof: The proof of Theorem 13 carries over in every detail.

Now suppose that A and B are sets having the same cardinality.[3] As in Theorem 17, we may consider groups $\mathscr{P}(A)$ and $\mathscr{P}(B)$ with composition of mappings as group operation in both cases. Intuitively it is plausible, exactly as in the discussion on page 82, that the groups $\mathscr{P}(A)$ and $\mathscr{P}(B)$ are algebraically indistinguishable, i.e., that they are *isomorphic*. This result will be deduced in Chapter 3.

DEFINITION 18.   The group $\mathscr{P}(A)$, with composition of mappings as group operation, is the *full transformation group on* A.

As in the discussion of the symmetric groups, it is customary to drop the symbol "$\circ$" as the notation for composition of mappings. Thus, if $\varphi, \psi \in \mathscr{P}(A)$, one writes "$\varphi\psi$" in place of "$\varphi \circ \psi$" to denote the composite of φ and ψ.

A number of definitions given in Section 6 are entirely appropriate for full transformation groups.

---

[3] Sets A and B *have the same cardinality* if and only if there is a one-one correspondence between A and B.

DEFINITION 19. Let $\varphi \in \mathscr{P}(A)$ and let $x, y \in A$. Then x is φ-*equivalent* to y, $x \sim_\varphi y$, if and only if there is an integer h such that $\varphi^h(x) = y$.

As in the restricted case of Section 6, $\sim_\varphi$ is an equivalence relation on A, and therefore it partitions A into equivalence classes.

DEFINITION 20. The equivalence classes determined by $\sim_\varphi$ are the *orbits* of φ. An orbit of φ is *trivial* if and only if it consists of a single element. Otherwise it is *nontrivial*. A permutation $\gamma \in \mathscr{P}(A)$ is a *cycle* if and only if it has at most one nontrivial orbit. The *displaced set*, D_φ, of a permutation φ is the the union of its nontrivial orbits. Two permutations φ, ψ are *disjoint* if and only if $D_\varphi \cap D_\psi = \phi$.

It is clear that each permutation φ completely determines its orbits. If γ is a cycle, then its nontrivial orbit is its displaced set.

Exercise: Let $\varphi, \psi \in \mathscr{P}(A)$. Prove that if φ and ψ are disjoint, then $\varphi\psi = \psi\varphi$.

We now inquire whether each permuation $\varphi \in \mathscr{P}(A)$ can be factored into a product of pairwise disjoint cycles. The example below shows that without some additional condition on φ, such a factorization is not possible.

EXAMPLE 14. Let $A = Z$ be the set of all integers and define $\varphi : Z \to Z$ by

$$\varphi(x) = -x,$$

for each $x \in Z$. Clearly φ is a one-one mapping of Z onto itself, therefore $\varphi \in \mathscr{P}(Z)$. From the definition of φ it follows that

$$(12) \qquad D_\varphi = Z - \{0\},$$

and so D_φ contains infinitely many elements. We prove:

If φ were a product of pairwise disjoint cycles $\gamma_1, \gamma_2, \ldots, \gamma_r, r \geq 1$, then D_φ would be finite. This result is obtained by showing that

$$(13) \qquad D_\varphi = D_{\gamma_1} \cup D_{\gamma_2} \cup \cdots \cup D_{\gamma_r},$$

and that each D_{γ_i} is finite.

We require the following simple facts which are left for the reader to verify: 1) If φ is a permutation, then $D_{\varphi^2} \subset D_\varphi$; 2) hence, if φ and ψ are disjoint permutations, so are φ^2 and ψ^2; 3) if γ is a cycle and D_γ has more than two elements, then γ^2 is not the identity.

Assuming that $\varphi = \gamma_1\gamma_2 \ldots \gamma_r, r \geq 1$, and the $\gamma_i, 1 \leq i \leq r$ are pairwise disjoint cycles, then equation (13) is obtained as in the proof of Exercise 8, page 79. Since the cycles γ_i are pairwise disjoint, and since $\varphi^2 = \iota$, the identity transformation,

$$(14) \qquad \iota = \varphi^2 = \gamma_1^2\gamma_2^2 \cdots \gamma_r^2,$$

and the $\gamma_i^2, 1 \leq i \leq r$, are also pairwise disjoint. We claim that each $\gamma_i^2 = \iota$,

$1 \leq i \leq r$. For, if $x \in Z$, then $x \in D_{\gamma_j^2}$ for at most one value of j. Therefore by the disjointness of the γ_i^2, $1 \leq i \leq r$,

$$x = \iota(x) = \gamma_j^2(x),$$

whence $\gamma_j^2 = \iota$, $j = 1$, 2, $\ldots$, r. Consequently $|D_{\gamma_j}| \leq 2$, $1 \leq j \leq r$, and this is the desired contradiction.

The following modification of Theorem 16 is valid for arbitrary sets A and is left as an exercise.

THEOREM 17.   Let $\varphi \in \mathscr{P}(A)$ be a permutation such that D_φ is a *finite set*. Then there exist cycles $\gamma_1, \gamma_2, \ldots, \gamma_r$ in $\mathscr{P}(A)$, $r \geq 1$, such that each pair $\gamma_i, \gamma_j, i \neq j$, is disjoint and such that

$$\varphi = \gamma_1 \gamma_2 \cdots \gamma_r.$$

Moreover, the cycles $\gamma_1, \gamma_2, \ldots, \gamma_r$ are uniquely determined up to order.

The proof is obtained by minor alterations in the proof of Theorem 16.

Note that if A is finite, then since $D_\varphi \subset A$, D_φ is also finite so that in this case the additional hypothesis—that D_φ is finite—is redundant.

9.   *Restrictions of Binary Operations*

In this section, we examine a technical point in connection with binary operations that is important for the definition and study of *subgroups*.

Let A, B and C be sets where $C \subset A$, and let $f : A \longrightarrow B$ be a mapping. The mapping

$$g : C \longrightarrow B$$

defined by

(15) $$g(x) = f(x), \quad x \in C,$$

is the *restriction of f to C*, written "$f|C$," and f is the *extension of g to A*. Thus

$$f | C = g,$$

where g is the function defined by (15).

The mappings that concern us here are the binary operations, namely, mappings of the form $* : S \times S \longrightarrow S$. If T is a subset of S, then $T \times T \subset S \times S$, and we set $\circledast = * | T \times T$, so that

$$\circledast : T \times T \longrightarrow S.$$

If $(a, b) \in T \times T$, then $(a, b) \in S \times S$, and therefore

$$a \circledast b = a * b.$$

The main point to be emphasized is this: Although the restriction $\circledast$ is clearly a mapping $T \times T \longrightarrow S$, it may *not* be a binary operation on T. Indeed, $\circledast$ will be a binary operation on T if and only if for each $(a, b) \in T \times T$,

$$a \circledast b = a * b \in T.$$

We illustrate with

EXAMPLE 15. Let Z be the set of all integers, and let $+$ be the addition of integers. Then

$$+ : Z \times Z \longrightarrow Z,$$

hence $+$ is a binary operation on Z. Next, let $\mathcal{O}$ be the set of all odd integers; then $\mathcal{O} \times \mathcal{O} \subset Z \times Z$. If $\oplus$ is the restriction of $+$ to $\mathcal{O} \times \mathcal{O}$, then

$$\oplus : \mathcal{O} \times \mathcal{O} \longrightarrow Z.$$

But $\oplus$ is *not* a binary operation on $\mathcal{O}$. For, if $(a, b) \in \mathcal{O} \times \mathcal{O}$, then $a \oplus b = a + b$ is an even integer, hence $a \oplus b \notin \mathcal{O}$. Therefore $\oplus$ is not a mapping $\mathcal{O} \times \mathcal{O} \longrightarrow \mathcal{O}$; hence it is not a binary operation on $\mathcal{O}$.

Some further examples are:

(a) Let Q be the set of all rational numbers, let Z be the set of all integers and let $\cdot$ be multiplication of rational numbers. The restriction of $\cdot$ to the integers is a binary operation on Z.

(b) Let Z be the set of all integers, let $\mathcal{O}$ be the set of all odd integers and let $\cdot$ be multiplication of integers. The restriction of $\cdot$ to $\mathcal{O}$ is a binary operation on $\mathcal{O}$.

(c) Let C be the set of all complex numbers and let $\mathcal{I}$ be the set of all pure imaginary numbers,

$$\mathcal{I} = \{i\,x \,|\, x \text{ real}\},$$

where $i^2 = -1$. Let $\cdot$ be multiplication of complex numbers. The restriction of $\cdot$ to $\mathcal{I}$ is *not* a binary operation on $\mathcal{I}$.

The next theorem provides a simple criterion for determining whether a restriction of a binary operation is a binary operation.

THEOREM 18. Let $*$ be a binary operation on a set S, let T be a subset of S and let $\circledast$ be the restriction of $*$ to $T \times T$. Then $\circledast$ is a binary operation on T if and only if for each $(a, b) \in T \times T$, $a * b \in T$.

Proof: First suppose $\circledast$ is a binary operation on T; then $a \circledast b \in T$ for each $(a, b) \in T \times T$. Since for each $(a, b) \in T \times T$, $a * b = a \circledast b$, it follows that $a * b \in T$.

　　　　Conversely, if for each $(a, b) \in T \times T$, $a * b \in T$, then $a \circledast b \in T$. Since the restriction of a mapping is a mapping, it follows that ⊛ is a mapping of $T \times T$ into T, i.e., ⊛ is a binary operation on T.

Q.E.D.

DEFINITION 21. Let S, T and $*$ be as in the statement of Theorem 18. T is *closed with respect to* $*$ if and only if for each $(a, b) \in T \times T$, $a * b \in T$.

In view of Theorem 18 we may state:
T is closed with respect to $*$ if and only if ⊛ is a binary operation on T.

Exercise: Let $*$ be a binary associative operation on a set S and let $T \subset S$. Prove: If T is closed with respect to $*$, then ⊛ $= * | T \times T$ is associative. Give an example to show that the closure hypothesis is needed.

10. *Subgroups*

We now consider the possibility that a "portion" of a group may itself be a group. The next two examples illustrate the ideas we have in mind.

EXAMPLE 16. Let Q and Z be the rational numbers and the integers, respectively. Further, let $+$ be the operation of addition of rationals and (for the purposes of this example) let $+'$ be the addition of integers. With respect to $+$, Q is a group and with respect to $+'$, Z is also a group. Indeed, the additive group of the integers is a *subgroup* of the additive group of the rationals. Note that $Z \subset Q$ and $+'$ is the restriction of $+$ to Z.

EXAMPLE 17. As a matter of notational convenience let the symmetric group $\mathscr{S}_4$ be the group of all permutations of the set $\{1, 2, 3, 4\}$. With this choice of the underlying set, the elements of $\mathscr{S}_4$ are represented simply as

$$(12), (23), (12)(34), (234), \text{ etc.}$$

Now let $\mathscr{F}$ be the subset

$$\mathscr{F} = \{(1), (12)(34), (14)(23), (13)(24)\}.$$

Define a mapping $\circ' : \mathscr{F} \times \mathscr{F} \longrightarrow \mathscr{F}$ by means of the table:

TABLE 4

$\circ'$	(1)	(12)(34)	(14)(23)	(13)(24)
(1)	(1)	(12)(34)	(14)(23)	(13)(24)
(12)(34)	(12)(34)	(1)	(13)(24)	(14)(23)
(14)(23)	(14)(23)	(13)(24)	(1)	(12)(34)
(13)(24)	(13)(24)	(14)(23)	(12)(34)	(1)

From the table it is clear that $\circ'$ is a binary operation on $\mathscr{F}$, that (1) is a neutral element for $\mathscr{F}$ and that each element in $\mathscr{F}$ has an inverse with respect to the binary operation $\circ'$. By direct computation, using the table, one can show that $\circ'$ is associative. Hence the structure consisting of $\mathscr{F}$ and the binary operation $\circ'$ is a group; this group is known as the "Four Group." The group $\mathscr{F}$ is a *subgroup* of the symmetric group $\mathscr{S}_4$. In this example we observe that $\mathscr{F} \subset \mathscr{S}_4$, and the reader may verify that $\circ'$ is the restriction of $\circ$ to $\mathscr{F}$.

With examples 16 and 17 as guides we are ready to formulate the definition of *subgroup*. Let G be a group with group operation $*$, let H be a subset of G and let $\circledast$ be the restriction of $*$ to $H \times H$.

DEFINITION 22.   The structure consisting of H and $\circledast$ is a *subgroup* of G if and only if H is a group with respect to $\circledast$ as group operation.

Exercise.   Let Q_+ be the set of positive rationals. With respect to multiplication, Q_+ is a group. Is it a subgroup of the additive group of the rationals? Why?

If the structure consisting of H and $\circledast$ is a subgroup of G we shall say, simply, "H is a subgroup of G" and write

$$H < G.$$

The subsets $H = \{e\}$ and $H = G$ are obviously subgroups of G; these are the *trivial* subgroups. All other subgroups are *proper*.

Now suppose $H < G$. Since H is a group, it has a neutral element e'.

THEOREM 19.   The neutral element, e', of H is also the neutral element of G.

Proof: Since e' is the neutral element of H, $e' \circledast e' = e'$. On the other hand, since $e' \in H$, also $e' \in G$ and $e' \circledast e' = e' * e'$ (why?). Hence

$$e' * e' = e'.$$

By Exercise 4, bottom of page 62, e' is the neutral element of G.
Q.E.D.

While the result stated in Theorem 19 contains no surprises, the example below shows that there exist simple group-like structures in which Theorem 19 is false.

EXAMPLE 18.   Let $G = \{0, 1, 2, 3, 4, 5\}$ and define $*$ on G as follows: If $x, y \in G$, then there is a unique pair of integers q and r such that

$$xy = 6q + r,$$

where $0 \le r < 6$. We then set

$$x * y = r.$$

It is easy to check that $*$ is a binary, associative operation on G and that 1 is a left neutral element for G. Of course, G together with $*$ do not constitute a group, since the element 0 has no (left) inverse in G. The system consisting of G and $*$ is a *semigroup*.

Now let $H = \{0, 2, 4\}$ and let $\circledast = * \mid H \times H$. The set H is closed with respect to $*$ and in fact—without going into details—H together with $\circledast$ is a *sub-semigroup* of G. However, as we can verify easily, this new system has no left neutral element, and therefore Theorem 19 is false in the present situation.

In practice, Definition 22 is a cumbersome criterion to use for determining whether or not a given group H is a subgroup of a group G. The next two theorems yield tests which are easier to apply.

THEOREM 20. H is a subgroup of G if and only if

(i) H is a nonempty subset of G,

(ii) H is closed with respect to $*$, and

(iii) if $x \in H$, then the inverse of x in G is also an element in H.

Proof: a) Let H be a subgroup of G. Then H is a group, and therefore $H \neq \emptyset$. If x and y are elements in H, then since H is a group with respect to the binary operation $\circledast$, $x \circledast y \in H$. But since $\circledast$ is the restriction of $*$ to $H \times H$, $x \circledast y = x * y$. Hence $x * y \in H$, and therefore H is closed with respect to $*$. Thus (ii) is proved.

Finally let $x \in H$; since H is a group, x has an inverse, y, in H. But since $x \in H$, x is also an element in G. Hence x has an inverse, z, in G. To prove (iii), we show that $y = z$. Since y is the inverse of x in H,

$$y \circledast x = e$$

where e is the neutral element of both H and G (Theorem 19). But, $y \circledast x = y * x$ (why?), so

$$y * x = e.$$

Hence y is the inverse of x in G. But x has exactly one inverse in G. Therefore $y = z$ and (iii) is proved.

b) Conversely, let H be a subset of G such that (i), (ii) and (iii) hold; we prove that $H < G$.

Since $H \neq \emptyset$, there is an $x \in H$; by (iii) $x^{-1} \in H$, and

$$x^{-1} * x = e.$$

But by (ii), $x^{-1} * x \in H$, hence the neutral element e of G is also an element in H. As usual, let $\circledast$ be the restriction of $*$ to $H \times H$. Then (ii) and Theorem 18 imply that $\circledast$ is a binary operation on H. By the exercise of page 87, $\circledast$ is associative. Further, for each $x \in H$,

$$e \circledast x = e * x = x,$$

so e is a left neutral element for H. Finally, if $x \in H$, then since $x^{-1} \in H$,

$$x^{-1} \circledast x = x^{-1} * x = e,$$

and thus x^{-1} is a left inverse of x in H. Therefore H is a group with respect to $\circledast$, and so $H < G$ by Definition 22.

Q.E.D.

THEOREM 21. H is a subgroup of G if and only if
(α) H is a nonempty subset of G, and
(β) for each x and for each y in H, $x^{-1} * y \in H$.

> *Proof:* a) Suppose $H < G$; then H is a group and therefore it is non-empty. If $x, y \in H$, then by (iii) of Theorem 20, x^{-1} and y are elements in H, hence by (ii) of Theorem 20, $x^{-1} * y \in H$.
>
> b) Conversely, suppose (α) and (β) hold. Since H is non-empty, there is an x in H. By (β), $e = x^{-1} * x \in H$. But, now since x and e are elements in H, $x^{-1} = x^{-1} * e \in H$. Thus we have shown that for each $x \in H$, x^{-1} is also in H.
>
> Finally, let x and y be elements in H. By the foregoing x^{-1} and y are in H, hence (by (β)), $(x^{-1})^{-1} * y \in H$. But since $(x^{-1})^{-1} = x$, $x * y \in H$, and so H is closed with respect to $*$. In short, we have proved that (α) and (β) imply conditions (i), (ii) and (iii) of Theorem 20. Therefore $H < G$.

Q.E.D.

Exercises

1. State and prove Theorems 20 and 21 for additive groups.
2. Let G be the group defined in Exercise 4, page 56. Verify that $H = \{(a, 0) \mid a \in Z\}$ and $K = \{(0, a) \mid a \in Z\}$ are subgroups of G.
3. List all the subgroups of $\mathscr{S}_4$. Which are cyclic? Which are commutative?
4. Let $x \in A_n = \{1, 2, \ldots, n\}$, $n \geq 1$, and let H be the subset of $\mathscr{S}_n$ defined by

$$H = \{\varphi \mid \varphi \in \mathscr{S}_n \text{ and } \varphi(x) = x\}.$$

Prove: a) H is a subgroup of $\mathscr{S}_n$ and b) if $n > 2$, then $\bigcap_{\varphi \in H} F_\varphi = \{x\}$, where $F_\varphi = \{y \mid y \in A_n \text{ and } \varphi(y) = y\}$.

5. Let B be a subset of $A_n = \{1, 2, \ldots, n\}$, $n \geq 1$, and let

$$K = \{\varphi \mid \varphi \in \mathscr{S}_n \text{ and } \varphi(x) = x \text{ for each } x \in B\}.$$

Prove: a) K is a subgroup of $\mathscr{S}_n$ and b) if $n > |B| + 1$, then $\bigcap_{\varphi \in K} F_\varphi = B$, where F_φ is defined as in Exercise 4.

6. Let A_n, B, K and $\mathscr{S}_n$ be as in Exercise 5, and set

$$C = \{y \mid y \in A_n \text{ and } \varphi(y) = y \text{ for each } \varphi \in K\}.$$

Prove that $B \subset C$. Give an example to show that $B = C$ cannot be deduced.

7. Let $K < H$ and $H < G$; prove that $K < G$.

8. Let $H < G$ and $K < G$. Prove: a) $H \cap K < H$; b) $H \cap K < K$; and c) $H \cap K < G$.

9. Let H, K and G be as in Exercise 8, and let $S < H$ and $S < K$. Prove that $S < H \cap K$. (By reason of Exercises 7 and 8, $H \cap K$ is said to be the "largest" subgroup contained in both H and K.)

10. Let H and K be subgroups of G. Prove that if $H \cup K < G$, then $H \subset K$ or $K \subset H$. Suppose H, K and L are subgroups of G such that $H \cup K \cup L < G$. What are the corresponding inclusion relationships among H, K and L?

11. Let $\mathscr{H}$ be a set of subgroups of a group G. Prove that $\cap \mathscr{H}$ is a subgroup. Further, show that if $H \in \mathscr{H}$, then $\cap \mathscr{H} < H$. Note that Exercise 11 is a generalization of Exercise 8.

11. A Discussion of Subgroups

Now that we have defined the subgroup concept and given simple criteria by which a subset can be recognized as a subgroup, we ask: "Given a group G, what can be said about its subgroups? More specifically, how many are there, and what are they like?" In general, and for arbitrary groups G, these questions are extremely difficult to answer. Indeed, complete answers have not yet been found, nor is it likely that they will be discovered in the near future. Some of the most difficult parts of group theory are concerned with answering different aspects of these questions. In view of all this we can at most give here some elementary, first steps in the direction of answers. Perhaps the best way to begin is by analyzing the pattern of subgroups of the most familiar of all groups, namely, the additive group Z of integers. From this analysis one may be led to further conjectures, some of which are presented as exercises at the end of this section.

EXAMPLE 19. We consider the subgroups of the additive group Z. First let us note that if n is a nonnegative integer, then the set $nZ = \{kn \mid k \in Z\}$ is a subgroup of Z (details?). We claim that every subgroup of Z is an nZ for some nonnegative integer n.

Let H be a subgroup of Z. If $H = \{0\}$, then $H = \{k \cdot 0 \mid k \in Z\} = 0Z$. Suppose $H \neq \{0\}$; then H contains a nonzero integer x and it also contains $-x$. Since one of the two integers, x, $-x$, is positive, we now know that H contains a positive integer. By the WOP, H contains a smallest positive integer m. If $m = 1$, then clearly H contains every integer and so $H = Z$, the other trivial subgroup. Suppose $m > 1$. Then since $-m \in H$,

$$2m = m - (-m) \in H \text{ and } -2m = -m - m \in H;$$

continuing in this way (more precisely, by the PFI), one sees that for each integer k

$$km \in H;$$

thus

(16) $\{km \,|\, k \in Z\} \subset H.$

We now claim that the inclusion (16) is, in fact, an equality. For, suppose $y \in H$. By the division algorithm, there exist integers q and r such that

$$y = qm + r,$$

where $0 \leq r < m$. By (16), $qm \in H$. Since $y \in H$,

$$r = y - qm \in H.$$

In view of the fact that m is the smallest positive integer in H and $r \in H$ where $0 \leq r < m$, it follows that $r = 0$. Hence $y = qm$ and so $H \subset \{km \,|\, k \in Z\}$. Consequently $H = \{km \,|\, k \in Z\}$.

In summary, we have proved:

H is a subgroup of Z if and only if there is a nonnegative integer m such that

$$H = \{km \,|\, k \in Z\} = mZ.$$

If $m = 0$, H is the trivial subgroup $\{0\}$; if $m = 1$, H is the trivial subgroup Z.

We can now answer the question, "How many subgroups are there?" in the case that $G = Z$, the additive group of integers. Our analysis has shown that for each nonnegative integer n, there is exactly one subgroup of Z, namely, nZ. Consequently, there is a one-one correspondence between the set N of nonnegative integers and the set of subgroups of Z.

What are the subgroups of Z like? For this, too, a simple answer is available. The group Z is cyclic with 1 as generator. Each of its subgroups nZ, $n \geq 0$, is likewise cyclic and it has $n = n \cdot 1$ as a generator. (What are all the generators of nZ?)

These answers lead one in a natural way to consider the same questions for all cyclic groups, finite or infinite. But here the reader must get to work; the answers are obtained by solving the problems below.

Exercises

1. Prove that every subgroup of a cyclic group is cyclic.

2. Let G be a cyclic group of finite order. Prove that if $H < G$, then $|H| \,|\, |G|$. (*Hint:* Use the division algorithm.)

3. Let G be a cyclic group of order n and let $k \,|\, n$. Prove that there is one and only one subgroup H of G such that $|H| = k$.

To what extent can the results obtained in the foregoing exercises be generalized to arbitary groups? Later we shall prove (Corollary 4 to Theorem 32) that if G is a finite group, not necessarily cyclic, and if $H < G$, then $|H| \,|\, |G|$. However, if $n = |G|$ and if $m \,|\, n$, then it may happen, in the noncyclic case, that G has distinct subgroups of order m. For example, the groups $\{(1), (12)(34), (13)(24), (14)(23)\}$ and $\{(1), (12), (34), (12)(34)\}$ are both subgroups of order four, of $\mathscr{S}_4$. This same example shows that one cannot expect all the subgroups of a *non*cyclic group to be cyclic. Indeed, it is reasonable to inquire whether or not a given group G always has cyclic subgroups. An answer to this question is available immediately. Let $x \in G$; then the subset

$$H = \{x^k \,|\, k \in Z\}$$

is easily seen to be a cyclic subgroup of G with generator x. Thus, for each element in G, there is a cyclic subgroup of G having the given element as generator.

Now suppose $x \in G$, and let (x) be the cyclic subgroup generated by x. Further, let

$$\mathscr{H} = \{H \,|\, H < G \text{ and } x \in H\}.$$

The reader can verify, without difficulty, that

(17) $$(x) = \bigcap \mathscr{H}.$$

The characterization of "cyclic subgroup generated by x" that equation (17) provides, enables us to generalize this concept to subgroups generated by sets of elements. To approach the idea gradually, let x and y be distinct elements in the group G and let

$$\mathscr{K} = \{K \,|\, K < G \text{ and } x \in K \text{ and } y \in K\}.$$

By Exercise 11, page 91, $\bigcap \mathscr{K}$ is a subgroup of G, and clearly $x \in \bigcap \mathscr{K}$ and $y \in \bigcap \mathscr{K}$. Next consider the set $[\{x, y\}]$ of all products

(18) $$a_1^{n_1} a_2^{n_2} \cdots a_r^{n_r},$$

where r is an integer, $r \geq 1$, each n_i is an integer, $1 \leq i \leq r$, and each $a_i \in \{x, y\}$, $1 \leq i \leq r$. We leave it to the reader to prove that $[\{x, y\}] < G$ and moreover that $[\{x, y\}] = \bigcap \mathscr{K}$. We define $[\{x, y\}]$ to be the *subgroup of G generated by the set $\{x, y\}$*.

The further extension of the concept of a *subgroup generated by a set*, as well as practice with this notion, is left to the reader in the next set of exercises.

Exercises

1. Let $\{(13), (1234)\}$ be a two-element subset of $\mathscr{S}_4$. (a) List all the subgroups of $\mathscr{S}_4$ containing both (13) and (1234). (b) List all the distinct products $a_1^{n_1} a_2^{n_2} \cdots a_r^{n_r}$ where each $a_i \in \{(13), (1234)\}$. (c) Do the same for the set $\{(123), (124)\}$.

2. Let S be a subset of a (multiplicative) group G. If $S = \emptyset$, set $[S] = \{e\}$; if $S \neq \emptyset$, let $[S]$ be the set of all products (18) where each $a_i \in S$, $1 \leq i \leq r$.

(a) Prove that [S] is a subgroup of G. (b) Let $\mathscr{H} = \{H \mid H < G$ and $S \subset H\}$. Prove that $[S] = \bigcap \mathscr{H}$. The subgroup $[S]$ is the *subgroup generated by the set S* and S is a *set of generators* of [S]. S is a set of generators of G if and only if $[S] = G$.

3. Prove that every subgroup H of a group G is generated by a subset of G.

4. In general, does a subgroup have only one set of generators?

5. Let $S = \{a, b\}$ be a set of generators for the group G, and suppose that $a^5 = e$, $b^2 = e$ and $ab = ba^4$. Construct a group table for G.

6. Let $H < G$ where $H \neq G$. Prove that the set $S = G - H$ (the complement of H relative to G) is a set of generators of G.

12. The Alternating Group

In the next few pages we establish a theorem which is of considerable value in the study of the symmetric group. In particular, this theorem is essential for our definition of a certain important subgroup of the symmetric group $\mathscr{S}_n$, namely, the *alternating group* $\mathscr{A}_n$.

The alternating group derives much of its original interest from its close connection with a historical problem of algebra. That problem was to determine whether the general equation of degree $n \geq 5$ is solvable by means of the arithmetic operations of addition, subtraction, multiplication and division, together with the extraction of roots, applied to the coefficients of the equation. In 1824 the young Norwegian mathematician Nils Henrik Abel proved that the general equation of degree n is not solvable in the manner described if $n \geq 5$. The method of proof depends upon establishing that the general equation of degree n is solvable if and only if the alternating group $\mathscr{A}_n$ has a certain property (also called "solvability"). Analysis shows that if $n \geq 5$, then $\mathscr{A}_n$ lacks this property and therefore the equation is not solvable in the desired way. The actual solution and discussion of the more general Galois Theory can be found in many texts (see, for example, I. N. Herstein, *Topics in Algebra*, chapter 5, New York, Blaisdell, 1964).

A simple way to define the alternating group is to utilize the result established in Exercise 3, page 82, namely, that every permutation is a product of transpositions. Now the aforementioned exercise shows at once that a permutation may be represented in different ways as products of transpositions. Thus, if 1, i and j are distinct, then

$$(x_i x_j) = (x_1 x_i)(x_1 x_j)(x_1 x_i).$$

However, the important feature of representations of permutations as products of transpositions is the invariance of *parity* (see the statement of Theorem 22, below), and it is this invariance which makes our definition of $\mathscr{A}_n$ work. The main theorem is

THEOREM 22.   Let

$$\varphi = \tau_1\tau_2 \cdots \tau_j = \tau_1'\tau_2' \cdots \tau_k',$$

$\varphi \in \mathscr{S}_n$, be two factorizations of φ into products of transpositions. Then the integers j and k have the same parity, i.e., both j and k are even, or else both are odd.

LEMMA.   If τ is a transposition in $\mathscr{S}_n$ and $\varphi \in \mathscr{S}_n$, then

$$N(\tau\varphi) = N(\varphi) \pm 1,$$

and

$$N(\varphi\tau) = N(\varphi) \pm 1,$$

where $N(\sigma)$ is the number of orbits of $\sigma \in \mathscr{S}_n$.

Proof: The proof is carried out only for the product $\tau\varphi$; the other half is left to the reader.

Let $\varphi = \gamma_1\gamma_2 \cdots \gamma_k$ be the factorization of φ into disjoint cycles and let $\tau = (ab)$. We shall see that if τ and φ are disjoint, then the proof of the lemma is quite direct. On the other hand, if τ and φ are not disjoint, then we must investigate a number of subcases.

CASE 1. τ and φ are disjoint. Then

$$\tau\varphi = \tau\gamma_1\gamma_2\cdots \gamma_k$$

is the factorization of $\tau\varphi$ into disjoint cycles. If m_i is the number of elements in the nontrivial orbit of γ_i, $1 \leq i \leq k$, then by Exercise 1, page 82.

$$\begin{aligned}N(\tau\varphi) &= (k + 1) + (n - (2 + m_1 + m_2 + \cdots + m_k)) \\ &= k + (n - (m_1 + m_2 + \cdots + m_k)) - 1 \\ &= N(\varphi) - 1.\end{aligned}$$

CASE 2. τ and φ are not disjoint.
(a) $\varphi = (a_1a_2 \cdots a_m)$, $m > 1$, is a cycle and $\tau = (ab)$ where $a =$ some a_i, $1 \leq i \leq m$, and $b \neq a_j$, $1 \leq j \leq m$. Without loss of generality we may assume $i = 1$ so that

$$\begin{aligned}\tau\varphi &= (a_1b)(a_1a_2 \cdots a_m) \\ &= (a_1a_2 \cdots a_mb).\end{aligned}$$

Hence

$$\begin{aligned}N(\tau\varphi) &= 1 + (n - (m + 1)) \\ &= 1 + (n - m) - 1 \\ &= N(\varphi) - 1.\end{aligned}$$

(b) $\varphi = (a_1a_2 \cdots a_m)$, $m > 1$, is a cycle and $\tau = (ab)$ where, say, $a = a_1$ and $b = a_j$, $j > 1$. If $j = 2$, then

$$\tau\varphi = (a_2a_3 \cdots a_m),$$

and

$$N(\tau\varphi) = 1 + (n - (m - 1)) = (1 + (n - m)) + 1 = N(\varphi) + 1.$$

Clearly the same result is obtained in case $j = m$. If $2 < j < m$, then

$$\tau\varphi = (a_1 a_j)(a_1 \cdots a_j \cdots a_m) = (a_1 \cdots a_{j-1})(a_j \cdots a_m),$$

and

$$\begin{aligned}N(\tau\varphi) &= 2 + n - ((j - 1) + (m - (j - 1)))\\ &= 2 + (n - m)\\ &= (1 + (n - m)) + 1\\ &= N(\varphi) + 1.\end{aligned}$$

(c) $\varphi = (a_1 a_2 \cdots a_{m_1})(b_1 b_2 \cdots b_{m_2})$, $m_1 > 1$ and $m_2 > 1$, is a product of disjoint cycles and $\tau = (a_1 b_1)$. Then we compute

$$\tau\varphi = (b_1 b_2 \cdots b_{m_2} a_1 \cdots a_{m_1}),$$

whence

$$N(\tau\varphi) = 1 + (n - (m_1 + m_2)).$$

But since $N(\varphi) = 2 + (n - (m_1 + m_2))$, we see that

$$N(\tau\varphi) = N(\varphi) - 1.$$

The remaining cases are easily derived from the foregoing and are left as exercises. The reader should make use of the fact that since the cycles $\gamma_1, \gamma_2, \ldots, \gamma_k$ in the factorization of φ are disjoint, they commute with each other. Hence one may always assume, in Case 2, that the orbits of τ and of the product $\gamma_3 \gamma_4 \cdots \gamma_k$ are disjoint.

Exercise: Complete the proof of the lemma.

We now turn to the

Proof of Theorem 22: We make use of the fact that two integers, j and k, have the same parity if and only if $(-1)^j = (-1)^k$. Define a mapping $\epsilon: \mathscr{S}_n \longrightarrow \{1, -1\}$ by

$$\epsilon(\varphi) = (-1)^{n-N(\varphi)}.$$

If ι is the identity permutation in $\mathscr{S}_n$, then

$$\epsilon(\iota) = (-1)^{n-N(\iota)} = (-1)^{n-n} = 1,$$

and for each transposition τ,

$$\epsilon(\tau) = (-1)^{n-N(\tau)} = (-1)^{n-(n-1)} = -1.$$

Moreover, if τ is a transposition and $\varphi \in \mathscr{S}_n$, then by the lemma

$$\epsilon(\tau\varphi) = (-1)^{n-N(\tau\varphi)} = (-1)^{n-(N(\varphi)\pm 1)} = -\epsilon(\varphi).$$

Applying the PFI to this last result, we deduce that

$$\epsilon(\tau_1 \tau_2 \cdots \tau_m) = (-1)^m.$$

Now suppose that

$$\varphi = \tau_1\tau_2 \cdots \tau_j = \tau_1'\tau_2' \cdots \tau_k'$$

are two factorizations of φ into products of transpositions. Then

$$\epsilon(\varphi) = (-1)^j = (-1)^k,$$

and consequently j and k have the same parity.

Q.E.D.

Theorem 22 justifies

DEFINITION 23. A permutation $\varphi \in \mathscr{S}_n$, $n \geq 2$, is *even* if and only if it is a product of an even number of transpositions; otherwise it is *odd*.

Let $\mathscr{A}_n$ be the set of all even permutations in $\mathscr{S}_n$, $n \geq 2$. Clearly a product of even permutations is even, and the inverse of an even permutation is likewise even. Therefore (by Theorem 20)

THEOREM 23. $\mathscr{A}_n$ is a subgroup of $\mathscr{S}_n$.

DEFINITION 24. For $n \geq 2$, $\mathscr{A}_n$ is the *alternating group*.

Exercises

1. List all the elements in $\mathscr{A}_2$, $\mathscr{A}_3$, and in $\mathscr{A}_4$.
2. Prove that $|\mathscr{A}_n| = \frac{1}{2}|\mathscr{S}_n| = n!/2$.
3. Prove that $\mathscr{A}_n$ is generated by the set of 3-cycles $\{(x_1x_ix_j) \mid 1, i, j$ mutually distinct$\}$.
4. Verify: If $n \geq 4$ and $1, i, j$ and k are mutually distinct, then $(x_ix_jx_k) = (x_1x_jx_i)(x_1x_kx_j)(x_1x_ix_j)$. By Exercise 3 it follows that if $n \geq 3$, then all 3-cycles are in $\mathscr{A}_n$.
5. Prove: If $1, 2, i$ and j are mutually distinct, then $(x_1x_ix_j) = (x_1x_2x_j)^{-1}(x_1x_2x_i)$ $(x_1x_2x_j)$. Hence if $n \geq 3$, $\mathscr{A}_n$ is generated by the set of 3-cycles $\{(x_1x_2x_k) \mid 3 \leq k \leq n\}$.

13. *The Congruence of Integers*

In his great work, *Disquisitiones Arithmeticae* (1801), C. F. Gauss exploited the concept of *congruence modulo an integer* to study divisibility properties of integers. Here we shall use the same concept to take a second look at the subgroups of the additive group of the integers. Later it will be seen that the notion of congruence modulo an integer can be generalized to arbitrary groups G, and that the generalized concept is intimately related to the subgroups of G.

DEFINITION 25. Let $n \in Z$. An integer a is *congruent modulo* n to an integer b if and only if there is an integer k such that $a - b = kn$. If a is congruent modulo n to b, one writes

$$a \equiv b \pmod{n}$$

and reads this expression, "a is congruent to b mod n." The integer n is the *modulus* of the congruence.

Note that $a \equiv b \pmod{n}$ if and only if $a \equiv b \pmod{(-n)}$; therefore we may restrict the moduli of congruences to nonnegative integers.

THEOREM 24. *Congruence modulo* n *is an equivalence relation such that if* $a \equiv b \pmod{n}$, *then for each* $x \in Z$,

(i) $x + a \equiv x + b \pmod{n}$, *and*
(ii) $xa \equiv xb \pmod{n}$.

> *Proof:* The verification that $\equiv \pmod{n}$ is an equivalence relation is very easy, and is left to the reader.
>
> Next, if $a \equiv b \pmod{n}$, then there is an integer k such that
>
> $$kn = a - b = (x + a) - (x + b),$$
>
> whence $x + a \equiv x + b \pmod{n}$. Similarly, from $kn = a - b$, one deduces $(xk)n = xa - xb$ so that $xa \equiv xb \pmod{n}$.
>
> Q.E.D.

Exercises

1. What is $\equiv \pmod{0}$?

2. Let a, b be integers, and suppose $a = q_1n + r_1$, $b = q_2n + r_2$ where q_i, r_i, $i = 1, 2$ are integers and $0 \leq r_1 < n$ and $0 \leq r_2 < n$. Prove that $a \equiv b \pmod{n}$ if and only if $r_1 = r_2$.

3. Give a complete description of the partition of Z determined by (i) $\equiv \pmod{1}$, (ii) $\equiv \pmod{2}$, (iii) $\equiv \pmod{3}$.

4. Let p be a prime number. Prove that if $xa \equiv xb \pmod{p}$ and $x \not\equiv 0 \pmod{p}$, then $a \equiv b \pmod{p}$.

5. Prove: If $a \equiv c \pmod{n}$ and $b \equiv d \pmod{n}$, then $a + b \equiv c + d \pmod{n}$.
In Exercises 6 and 7, a_i, m_i, $i = 1, 2$, are integers, and m_1, m_2 are relatively prime.

• 6. Prove there is an integer u such that $u \equiv 1 \pmod{m_1}$ and $u \equiv 0 \pmod{m_2}$. (*Hint:* Since m_1, m_2 are relatively prime, there exist integers s and t such that $sm_1 + tm_2 = 1$.)

7. Deduce from Exercise 6 that there exist integers y and z such that $y \equiv a_1 \pmod{m_1}$ and $y \equiv 0 \pmod{m_2}$, and $z \equiv 0 \pmod{m_1}$ and $z \equiv a_2 \pmod{m_2}$. Hence show that there is an integer x such that

$$x \equiv a_1 \pmod{m_1} \text{ and } x \equiv a_2 \pmod{m_2}.$$

8. (Chinese Remainder Theorem). Let $m_1, m_2, \ldots, m_n$ be integers which are relatively prime in pairs and let $a_1, a_2, \ldots, a_n \in Z$. Prove that there exists $x \in Z$ such that

$$x \equiv a_i \,(\mathrm{mod}\ m_i), \quad 1 \leq i \leq m.$$

The connection between congruence *mod n* and the subgroups of Z is established painlessly. By definition, $a \equiv b \,(\mathrm{mod}\ n)$ if and only if $a - b = kn \in nZ$. In particular, $a \equiv 0 \,(\mathrm{mod}\ n)$ if and only if $a = kn \in nZ$. But, for each integer n, nZ is a subgroup of Z. Hence the set $\{a \mid a \in Z$ and $a \equiv 0 \,(\mathrm{mod}\ n)\}$ is the subgroup nZ of Z.

Conversely, if $H < Z$, then there is an integer $m \geq 0$ such that $H = mZ$. An integer a is in H if and only if $a = qm$ for some $q \in Z$; therefore $a \in H$ if and only if $a \equiv 0 \,(\mathrm{mod}\ m)$. Finally, $\equiv \,(\mathrm{mod}\ m)$ is the same as $\equiv \,(\mathrm{mod}\ n)$ if and only if $m = n$; also $mZ = nZ, m, n \in N$, if and only if $m = n$. Therefore we have proved:

There exists a one-one correspondence between the set of congruences, $\{\equiv \,(mod\ n) \mid n \in N\}$ *and the set* $\{nZ \mid n \in N\}$ *of all subgroups of Z. For each* $n \in N$,

$$nZ = \{a \mid a \in Z \text{ and } a \equiv 0 \,(\mathrm{mod}\ n)\}.$$

There are equivalence relations on Z which are not congruences. Indeed, since each partition of Z determines an equivalence relation on Z, all one need do is to construct an appropriate partition and from it obtain the equivalence relation it determines. Thus, for example, the set $\{A, B, C\}$, where

$$A = \{0\}, \qquad B = \{1, -1\}, \qquad C = Z - A \cup B$$

is a partition of Z (why?), but the equivalence relation it determines is not a congruence mod n (details?).

Among all equivalence relations on Z, how can we select precisely those which are congruences modulo an integer? We shall see that the key to the answer is condition (i) of Theorem 24.

DEFINITION 26. An equivalence relation $\sim$ on Z is *stable with respect to* $+$ (or simply, *stable*) if and only if $a \sim b$ implies $x + a \sim x + b$ for each $x \in Z$.

By Theorem 24, every congruence mod n is stable. On the other hand, the equivalence relation determined by the partition $\{A, B, C\}$, above, is not stable.

THEOREM 25. An equivalence relation on Z is a congruence mod n, for some integer n, if and only if it is stable.

Proof: As remarked above, $\equiv \,(\mathrm{mod}\ n)$ is stable.

Conversely, let $\sim$ be a stable equivalence relation on Z and set

$$H = \{x \mid x \in Z \text{ and } x \sim 0\};$$

we claim that $H < Z$. Since $0 \in H$, $H \neq \emptyset$. If $b \in H$, then $b \sim 0$ and by stability $(-b) + b \sim (-b) + 0 = -b$. Thus $0 \sim -b$ and so $-b \sim 0$, whence $-b \in H$. If $a \in H$, then $a \sim 0$ and $a + b \sim 0 + b = b$. But since $b \sim 0$, we deduce $a + b \sim 0$ and therefore $a + b \in H$. By Theorem 20, $H < Z$. Consequently there is an $n \in N$ such that $H = nZ$.

Now suppose $a \sim b$; stability implies that $a - b = a + (-b) \sim b + (-b) = 0$, i.e., $a - b \sim 0$. Therefore $a - b \in H = nZ$ and so $a \equiv b \pmod{n}$. Since the foregoing steps are reversible it follows that if $a \equiv b \pmod{n}$, then $a \sim b$. Consequently $\sim$ is $\equiv \pmod{n}$.

Q.E.D.

14. The Modular Arithmetics

The equivalence relations $\equiv \pmod{n}$ will now be used to construct new groups, the *modular arithmetics*, from the additive group of integers. The process to be described here will be generalized to arbitrary groups, and we shall see that the generalization has far-reaching consequences for the study of group theory.

Let $\mathscr{P}$ be the partition of Z determined by $\equiv \pmod{n}$, and let $\bar{a}$ be the equivalence class containing the integer a. Thus

$$\bar{a} = \{x \mid x \in Z \text{ and } x \equiv a \pmod{n}\}.$$

If $\bar{b}$ is the equivalence class containing b, then

(19) $\bar{a} = \bar{b}$ if and only if $a \equiv b \pmod{n}$.

Now let

$$a = q_1 n + r_1, \ 0 \leq r_1 < n, \text{ and } b = q_2 n + r_2, \ 0 \leq r_2 < n.$$

By Exercise 2, page 98, $a \equiv b \pmod{n}$ if and only if the remainders r_1 and r_2 are equal. Consequently

$$\bar{a} = \bar{b} \text{ if and only if } r_1 = r_2.$$

This fact inspires

DEFINITION 27.   The equivalence classes determined by $\equiv \pmod{n}$ are *residue* (or, *remainder*) classes mod n.

We claim that the residue classes $\bar{0}, \bar{1}, \ldots, \overline{n-1}$ are mutually distinct and moreover that

(20) $\mathscr{P} = \{\bar{0}, \bar{1}, \ldots, \overline{n-1}\}.$

For, if $\bar{a} = \bar{b}$ where $0 \leq a \leq n - 1$ and $0 \leq b \leq n - 1$, then $a \equiv b \pmod{n}$, so $a - b = kn$ for some integer k. If $n = 0$, then obviously $a = b$. If $n \neq 0$, then $n \,|\, a - b$. But since $0 \leq |a - b| \leq n - 1$, it follows that $a - b = 0$, hence again $a = b$. Therefore the residue classes $\bar{0}, \bar{1}, \ldots, \overline{n - 1}$ are mutually distinct.

To verify (20), let $\bar{u} \in \mathscr{P}$ where u is an integer; we show that $\bar{u}$ is one of the n residue classes $\bar{0}, \bar{1}, \ldots, \overline{n - 1}$. By the division algorithm, there exist integers q and r such that

$$u = qn + r,$$

where $0 \leq r < n$. Hence $u - r = qn$ and so $u \equiv r \pmod{n}$. By (19)

$$\bar{u} = \bar{r},$$

where r is one of the integers $0, 1, \ldots, n - 1$.

Exercise. If $n = 5$, what is the residue class containing (a) 43, (b) 7345, (c) 9?

We are going to make $\mathscr{P}$ into a group, and for this purpose we need a binary operation on $\mathscr{P}$. As a first step we require

THEOREM 26.   If $\bar{a} = \bar{c}$ and $\bar{b} = \bar{d}$, then $\overline{a + b} = \overline{c + d}$.

Proof: Use Exercise 5, page 98.

                                                           Q.E.D.

THEOREM 27. $\mathscr{P}$ is an Abelian group with respect to $\oplus$ as group operation, where $\oplus$ is defined by

(21) $$\bar{a} \oplus \bar{b} = \overline{a + b}.$$

Proof: It is an immediate consequence of Theorem 25 that $\oplus$ is a binary operation on $\mathscr{P}$. Next, $\oplus$ is associative. For,

$$\begin{aligned}
\bar{a} \oplus (\bar{b} \oplus \bar{c}) &= \bar{a} \oplus (\overline{b + c}) \\
&= \overline{a + (b + c)} \\
&= \overline{(a + b) + c} \quad \text{(by associativity of $+$ in Z)} \\
&= (\bar{a} \oplus \bar{b}) \oplus \bar{c}.
\end{aligned}$$

Further, $\bar{0}$ is a left neutral element since

$$\bar{0} \oplus \bar{a} = \overline{0 + a} = \bar{a}.$$

If $\bar{a} \in \mathscr{P}$, then $\overline{n - a} \in \mathscr{P}$ and

$$\overline{n - a} \oplus \bar{a} = \overline{(n - a) + a} = \bar{n} = \bar{0},$$

so that each $\bar{a} \in \mathscr{P}$ has a left inverse in $\mathscr{P}$. Finally,

$$\bar{a} \oplus \bar{b} = \overline{a + b} = \overline{b + a} = \bar{b} \oplus \bar{a},$$

and therefore $\oplus$ is commutative.

                                                            Q.E.D.

DEFINITION 28. The group $\mathscr{P}$ is the *modular arithmetic* with *modulus n*.

Exercises

Give detailed descriptions of the following modular arithmetics:
1. with modulus 5;
2. with modulus 0;
3. with modulus 1.

In Section 13, we found that the equivalence relation $\equiv$ (mod n) determines and is determined by the subgroup nZ. It is therefore pertinent to inquire if the modular arithmetic with modulus n can be obtained from the subgroup nZ without the intervention of the relation $\equiv$ (mod n). In order to answer this question, let us define:
If $A \subset Z$, and if $x \in Z$, then

$$x + A = \{x + y \mid y \in A\}.$$

Now let nZ be the subgroup of Z determined by $\equiv$ (mod n). Then

$$\begin{aligned}
0 + nZ &= \{0 + nk \mid k \in Z\} \\
&= \{nk \mid k \in Z\} \\
&= \{x \mid x \in Z \text{ and } x \equiv 0 \ (\text{mod } n)\} \\
&= \bar{0};
\end{aligned}$$

and similarly,

$$1 + nZ = \{1 + nk \mid k \in Z\} = \{x \mid x \in Z \text{ and } x \equiv 1 \ (\text{mod } n)\} = \bar{1}.$$

In general, for each $u \in Z$,

(22) $\bar{u} = u + nZ,$ (details?)

and so the elements of $\mathscr{P}$ are

$$\bar{0} = nZ, \ \bar{1} = 1 + nZ, \ldots, \overline{n-1} = (n-1) + nZ.$$

Thus the elements of $\mathscr{P}$ can be obtained directly from the subgroup nZ. As for the binary operation $\oplus$ on $\mathscr{P}$, this too can be obtained from nZ. Indeed, (21) and (22) show that

$$(a + nZ) \oplus (b + nZ) = (a + b) + nZ.$$

To illustrate the foregoing, the modular arithmetic with modulus 3 consists of residue classes

$$\bar{0} = 3Z, \quad \bar{1} = 1 + 3Z, \quad \bar{2} = 2 + 3Z,$$

and the binary operation $\oplus$ is given by the table below:

TABLE 5

$\oplus$	$\bar{0}$	$\bar{1}$	$\bar{2}$
$\bar{0}$	$\bar{0}$	$\bar{1}$	$\bar{2}$
$\bar{1}$	$\bar{1}$	$\bar{2}$	$\bar{0}$
$\bar{2}$	$\bar{2}$	$\bar{0}$	$\bar{1}$

It is customary to denote the modular arithmetic with modulus n by

$$Z/nZ,$$

or more simply by

$$Z_n.$$

These are the notations which we use henceforth.

15. *Equivalence Relations and Subgroups*

As we attempt to extend the concept of *congruence modulo n* to arbitrary groups, let us recall that in the additive group of integers $\equiv$ (mod n) was completely characterized as a stable equivalence relation (Theorem 25). This result suggests that for groups in general, we investigate equivalence relations which are stable with respect to the group operation. Now suppose G is a multiplicative, nonabelian group and that $\sim$ is an equivalence relation on G. It may happen that if $a \sim b$, then for each $x \in G$, $xa \sim xb$; but since G is nonabelian, there may be a $y \in G$ such that $ay \not\sim by$. Similarly, a situation may arise in which: If $a \sim b$, then for each $x \in G$, $ax \sim bx$; but there may be a $y \in G$ such that $ya \not\sim yb$. If such equivalence relations do occur, then the extensions of $\equiv$ (mod n) to arbitrary groups will require a small subtlety. Our next example shows that even for the simplest nonabelian groups there are asymmetrically stable equivalence relations.

EXAMPLE 20. Let $G = \mathscr{S}_3$ be the symmetric group on three elements.
(a) Define $\sim_\ell$ on $\mathscr{S}_3$ by:

$(1) \sim_\ell (1),\ (1) \sim_\ell (23),\ (23) \sim_\ell (1),\ (23) \sim_\ell (23),$

$(12) \sim_\ell (12),\ (12) \sim_\ell (123),\ (123) \sim_\ell (12),\ (123) \sim_\ell (123),$

$(13) \sim_\ell (13),\ (13) \sim_\ell (132),\ (132) \sim_\ell (13),\ (132) \sim_\ell (132).$

Although it is a tedious job, it can be verified that $\sim_\ell$ is an equivalence relation such that if $\varphi \sim_\ell \psi$, then for each $\eta \in \mathscr{S}_3$, $\eta\varphi \sim_\ell \eta\psi$. Thus we might say that $\sim_\ell$ is *left stable*. On the other hand, $(12) \sim_\ell (123)$ but $(12)(12) \not\sim_\ell$ $(123)(12)$, i.e., $\sim_\ell$ is not *right stable*.

(b) Define $\sim_r$ on $\mathscr{S}_3$ by:

$$(1) \sim_r (1),\ (1) \sim_r (23),\ (23) \sim_r (1),\ (23) \sim_r (23),$$

$$(12) \sim_r (12),\ (12) \sim_r (132),\ (132) \sim_r (12),\ (132) \sim_r (132),$$

$$(13) \sim_r (13),\ (13) \sim_r (123),\ (123) \sim_r (13),\ (123) \sim_r (123).$$

In this case it can be shown that if $\varphi \sim_r \psi$, then for each $\eta \in \mathscr{S}_3$, $\varphi\eta \sim_r \psi\eta$, i.e., $\sim_r$ is *right stable*. But, although $(12) \sim_r (132)$, $(12)(12) \not\sim_r (12)(132)$, so that $\sim_r$ is not *left stable*.

The foregoing considerations lead us to formulate

DEFINITION 29.   Let G be a multiplicative group, and let $\sim$ be an equivalence relation on the set comprising G. The relation $\sim$ is *left stable* if and only if

$$a \sim b \text{ implies } xa \sim xb \text{ for each } x \in G;$$

$\sim$ is *right stable* if and only if

$$a \sim b \text{ implies } ax \sim bx \text{ for each } x \in G;$$

finally, $\sim$ is *stable* if and only if it is both left and right stable.

If G is an additive group, the condition for left stability is: $a \sim b$ implies $x + a \sim x + b$ for each $x \in G$. Similarly, the condition for right stability is: $a \sim b$ implies $a + x \sim b + x$ for each $x \in G$. In an Abelian group, every left (right) stable equivalence relation is stable.

Continuing with Example 20, we now exhibit a stable equivalence relation on $\mathscr{S}_3$.

EXAMPLE 20 (c)   Define $\sim$ on $\mathscr{S}_3$ by:

$$(1) \sim (1),\ (1) \sim (123),\ (123) \sim (1),\ (123) \sim (123),$$

$$(1) \sim (132),\ (132) \sim (1),\ (132) \sim (123),\ (123) \sim (132),$$

$$(132) \sim (132),\ (12) \sim (12),\ (12) \sim (23),\ (23) \sim (12),$$

$$(23) \sim (23),\ (12) \sim (13),\ (13) \sim (12),\ (13) \sim (23),$$

$$(23) \sim (13),\ (13) \sim (13).$$

It can be verified that $\sim$ is both left and right stable, therefore it is stable.

Note that the simplest stable equivalence relation on G is equality.

The symmetry between left and right stability is complete; for each definition and/or theorem associated with left (right) stability there is a corresponding definition and/or theorem for right (left) stability. This being the case, we shall develop the theories of the two kinds of stability parallel to each other. The present developments will mimic the investigations in Section 13 closely.

Let G be a group, let $\sim$ be a left (right) stable equivalence relation on G, and let $\mathscr{P}$ be the partition of G determined by $\sim$. If A is an equivalence class of $\sim$, i.e., if $A \in \mathscr{P}$, and if $a \in A$, then by the definition of partition,

(22) $$A = \{x \mid x \in G \text{ and } x \sim a\}.$$

Further, $\mathscr{P}$ has the properties

(i) $\bigcup \mathscr{P} = G$;
(ii) if $A \in \mathscr{P}$, then $A \neq \emptyset$; and
(iii) if $A, B \in \mathscr{P}$, then $A = B$ or else $A \cap B = \emptyset$.

Now let e be the neutral element of G; by (i), there is an equivalence class $H \in \mathscr{P}$ such that $e \in H$, and by (22),

(23) $$H = \{x \mid x \in G \text{ and } x \sim e\}.$$

For instance, in Example 20, parts (a) and (b), the equivalence class $H \in \mathscr{P}$, containing e, is

$$H = \{(1), (23)\},$$

and obviously $H < \mathscr{S}_3$. This leads us to conjecture

THEOREM 28. (i) If $\sim$ is left (right) stable, and if $\mathscr{P}$ is the partition of G determined by $\sim$, then the equivalence class $H \in \mathscr{P}$ defined by equation (23), is a subgroup of G.
(ii) If $\sim$ is left stable, then $x \sim y$ if and only if $x^{-1}y \in H$; if $\sim$ is right stable, then $x \sim y$ if and only if $yx^{-1} \in H$.

Proof: We consider only the case that $\sim$ is left stable; the corresponding proof for right stability is left to the reader.
(i) Let $a, b \in H$; it suffices to show that $a^{-1}b \in H$. Since $a \in H$, $a \sim e$, and as $\sim$ is left stable, $e = a^{-1}a \sim a^{-1}e = a^{-1}$; hence $a^{-1} \sim e$. Further, since $b \sim e$, $a^{-1}b \sim a^{-1}e = a^{-1}$, whence by the transitivity of $\sim$, $a^{-1}b \sim e$. Therefore $a^{-1}b \in H$ and so $H < G$.
(ii) If $x \sim y$, then by left stability $e = x^{-1}x \sim x^{-1}y$, hence $x^{-1}y \sim e$ and so $x^{-1}y \in H$. Since each of these steps is reversible, the converse is obtained at once.

Q.E.D.

Thus, beginning with a left (right) stable equivalence relation $\sim$ on G, one obtains a subgroup H of G defined by equation (23).
Conversely, suppose a subgroup H of G is given. Does H determine a left and a right stable equivalence relation on G? The answer is "yes," and is obtained as follows:
For each x and each y in G, define

(24) $$x \sim' y \text{ if and only if } x^{-1}y \in H.$$

We claim that $\sim'$ is a left stable equivalence relation on G.

First, $x \sim' x$, for each $x \in G$. For, since $e \in H$, $x^{-1}x = e \in H$, and so by (24), $x \sim' x$. Next, $\sim'$ is symmetric. Indeed, if $x \sim' y$, then $x^{-1}y \in H$; but since H is a group, $y^{-1}x = (x^{-1}y)^{-1} \in H$, therefore $y \sim' x$. Moreover, $\sim'$ is transitive. Suppose $x \sim' y$ and $y \sim' z$; then $x^{-1}y$ and $y^{-1}z$ are in H, hence $(x^{-1}y)(y^{-1}z) = x^{-1}z \in H$, and so $x \sim' z$. Finally, $\sim'$ is left stable. For, if $x \sim' y$, and if $z \in G$, then $(zx)^{-1}(zy) = x^{-1}y \in H$. Hence, by (24), $zx \sim' zy$.

If in place of (24) we take

(24') $x \sim' y$ if and only if $yx^{-1} \in H$

as our definition, then $\sim'$ is a right stable equivalence relation on G. (Prove it!)

Now suppose a left (right) stable equivalence relation $\sim$ is given on G. Then equation (23) defines a subgroup H of G, and H together with (24) (alternatively, (24')) defines a left (right) stable equivalence relation $\sim'$ on G;

$$\text{left stable } \sim \xrightarrow{(23)} \text{subgroup } H \xrightarrow{(24)} \text{left stable } \sim',$$

and

$$\text{right stable } \sim \xrightarrow{(23)} \text{subgroup } H \xrightarrow{(24')} \text{right stable } \sim'.$$

In both cases, we have

THEOREM 29. $\sim = \sim'$.

> *Proof:* Suppose $x \sim y$ where $\sim$ is left stable. By Theorem 28, part (ii), $x^{-1}y \in H$ and by (24) $x \sim' y$. Since these steps are reversible, we have $x \sim y$ if and only if $x \sim' y$, hence $\sim = \sim'$. The case of right stability is left to the reader.
>
> Q.E.D.

In a similar fashion we may begin with a subgroup H of G, define an equivalence relation $\sim$ on G by means of (24) (by (24')), and then use (23) to define a subgroup H' of G; thus

$$\text{subgroup } H \xrightarrow{(24)} \text{left stable } \sim \xrightarrow{(23)} \text{subgroup } H',$$

and

$$\text{subgroup } H \xrightarrow{(24')} \text{right stable } \sim \xrightarrow{(23)} \text{subgroup } H'.$$

In this situation we have a result corresponding to Theorem 29, namely,

THEOREM 30. $H = H'$.

> *Proof:* Exercise.

The effect of Theorems 29 and 30 is to show that we may move freely from left (right) stable equivalence relation to subgroup by means of (23), and vice versa by means of (24) and (24'), and the original relation or sub-group will be recovered. Now if H is a subgroup of G, then H determines a left stable equivalence relation, say, $\sim_\ell$, by means of (24), and a right stable equivalence relation $\sim_r$ by means of (24'). These, in turn, determine partitions $\mathscr{P}_\ell$ and $\mathscr{P}_r$, respectively.

DEFINITION 30.   The partition $\mathscr{P}_\ell(\mathscr{P}_r)$ is the *left (right) coset decomposition of G determined by H*. Each equivalence class in $\mathscr{P}_\ell$ (in $\mathscr{P}_r$) is a *left (right) coset* of H.

Parts (a) and (b) of Example 20 show that in general we cannot expect $\mathscr{P}_\ell$ and $\mathscr{P}_r$ to be equal. Those subgroups H for which $\mathscr{P}_r = \mathscr{P}_\ell$ are of special interest and will be discussed in Section 17. Of course, if G is Abelian, then automatically $\mathscr{P}_r = \mathscr{P}_\ell$ and, in particular, if G is a modular arithmetic with modulus n (Section 14),

$$\mathscr{P}_r = \mathscr{P}_\ell = \{\bar{0}, \bar{1}, \ldots, \overline{n-1}\}.$$

Before proceeding with an example of a coset decomposition in the case that G is nonabelian, we introduce a convenient notational device: If S is a subset of a (multiplicative) group G, set

$$xS = \{xy \,|\, y \in S\} \text{ and } Sx = \{yx \,|\, y \in S\}.$$

(The corresponding definitions, in case G is additive, are

$$x + S = \{x + y \,|\, y \in S\} \text{ and } S + x = \{y + x \,|\, y \in S\}.)$$

In particular, S may be a subgroup of G. Thus if $G = \mathscr{S}_3$, if $S = \{(1), (23)\}$ and if $x = (12)$, then $(12)S = \{(12), (123)\}$ and $S(12) = \{(12), (132)\}$.

EXAMPLE 21.   We are going to discuss the left coset decomposition of $\mathscr{S}_3$ determined by the subgroup $H = \{(1), (12)\}$. According to (24) we define: If $\varphi,\ \psi \in \mathscr{S}_3$, then $\varphi \sim_\ell \psi$ if and only if $\varphi^{-1}\psi \in H$.

Then, $\sim_\ell$ is a left stable equivalence relation on $\mathscr{S}_3$, hence it determines a left coset decomposition, $\mathscr{P}_\ell$, of $\mathscr{S}_3$. If $\varphi \in \mathscr{S}_3$, there is an $X \in \mathscr{P}_\ell$ such that $\varphi \in X$, and moreover

$$X = \{\psi \,|\, \psi \in \mathscr{S}_3 \text{ and } \psi \sim_\ell \varphi\}.$$

In particular, there is a $Y \in \mathscr{P}_\ell$ such that the identity permutation $(1) \in Y$. And $\varphi \in Y$ if and only if $\varphi^{-1} = \varphi^{-1}(1) \in H$, hence if and only if $\varphi \in H$. Therefore $Y = H$. Next, there is an equivalence class $A \in \mathscr{P}_\ell$ such that $(13) \in A$ and so

$$A = \{\psi \,|\, \psi \in \mathscr{S}_3 \text{ and } \psi \sim_\ell (13)\}.$$

Consequently, $\psi \in A$ if and only if $\psi^{-1}(13) = (1)$ or $\psi^{-1}(13) = (12)$. In the former case $\psi = (13)$ and in the latter $\psi = (123)$. Thus $A = \{(13), (123)\}$ and so

$$A = (13)\{(1), (12)\} = (13)H.$$

By the same reasoning, there is a $B \in \mathscr{P}_\ell$ such that

$$B = (23)\{(1), (12)\} = (23)H.$$

Since $H \cup (13)H \cup (23)H = \mathscr{S}_3$ and $H \cap (13)H = H \cap (23)H = (13)H \cap (23)H = \emptyset$, it follows that

$$\mathscr{P}_\ell = \{H, (13)H, (23)H\}.$$

It is now reasonable to inquire, "What about the subsets $(12)H$, $(123)H$ and $(132)H$? Are they also equivalence classes in $\mathscr{P}_\ell$?" Indeed they are! In fact, one can check, by direct computation, that $(12)H = H$, $(123)H = (13)H$ and $(132)H = (23)H$. Thus for each equivalence class A in $\mathscr{P}_\ell$, there is an element $\varphi \in \mathscr{S}_3$ such that

$$A = \varphi H;$$

conversely, every such subset of $\mathscr{S}_3$, i.e., every subset φH where $\varphi \in \mathscr{S}_3$, is an element in $\mathscr{P}_\ell$. Therefore the left cosets of $\mathscr{S}_3$ determined by H are the subsets φH of $\mathscr{S}_3$ where $\varphi \in \mathscr{S}_3$. Moreover, if φH and ψH are left cosets determined by H, one can verify that

$$\varphi H = \psi H \text{ if and only if } \psi^{-1}\varphi \in H.$$

It will be very instructive for the reader to check the last assertion for the cosets H, $(12)H$, $(13)H$, $(23)H$, $(123)H$ and $(132)H$.

Exercises

1. Let A and B be subsets of a group G, and let $x, y, z \in G$. Prove: a) $xy \in xA$ if and only if $y \in A$; b) $A \subset B$ if and only if $xA \subset xB$ (if and only if $Ax \subset Bx$); c) for each x and y in G, $(xy)A = x(yA)$, $A(xy) = (Ax)y$ and $(xA)y = x(Ay)$.

2. Prove: If $H < G$, then $zH = H$ if and only if $z \in H$. Similarly, $Hz = H$ if and only if $z \in H$.

3. Carry out the details of Exercises 1 and 2 in the case that G is an additive group.

4. As in Example 21, let $H = \{(1), (12)\}$. Discuss the right coset decomposition of $\mathscr{S}_3$ determined by H.

5. Let $K = \{(1), (12), (34), (12)(34)\}$. Discuss the right and left coset decompositions of $\mathscr{S}_4$ determined by the subgroup K.

6. Let $\mathscr{F} = \{(1), (12)(34), (13)(24), (14)(23)\}$. Discuss both the left and the right coset decompositions of $\mathscr{S}_4$ determined by $\mathscr{F}$.

The final theorem of this section formalizes our experiences with Example 21 and Exercises 4, 5 and 6 above.

THEOREM 31. Let $H < G$ and let $\mathscr{P}_\ell$ and $\mathscr{P}_r$, respectively, be the left and right coset decompositions of G determined by H. Then

(i) for each $A \in \mathscr{P}_\ell$, there is an $x \in G$ such that $A = xH$; for each $A \in \mathscr{P}_r$, there is an $x \in G$ such that $A = Hx$;

(ii) for each $y \in G$, $yH \in \mathscr{P}_\ell$ and $Hy \in \mathscr{P}_r$;

(iii) $xH = yH$ if and only if $x^{-1}y \in H$; $Hx = Hy$ if and only if $yx^{-1} \in H$.

Proof: We prove the first half of each item and leave the remaining halves to the reader.

(i) Let $x \in A$; then $A = \{g \,|\, g \in G$ and $g \sim x\}$. If $z \in A$, then $z \sim x$ whence $x \sim z$ and ((ii) of Theorem 28) $x^{-1}z \in H$. Therefore $z = x(x^{-1}z) \in xH$ and so $A \subset xH$. On the other hand, if $z \in xH$, then $x^{-1}z \in x^{-1}(xH) = H$, hence $z \sim x$ and therefore $z \in A$. Consequently $xH \subset A$ and (i) is proved.

(ii) First note that for each $y \in G$, $y \in yH$ (why?). To prove that $yH \in \mathscr{P}_\ell$ it suffices to show that

$$yH = \{g \,|\, g \in G \text{ and } g \sim y\}.$$

If $z \in yH$, then $y^{-1}z \in H$ whence $z \sim y$. Therefore, $z \in \{g \,|\, g \in G$ and $g \sim y\}$. Since these steps are all reversible, (ii) is proved.

(iii) If $xH = yH$, then $x^{-1}(xH) = x^{-1}(yH)$ or $H = (x^{-1}y)H$. By Exercise 2, above, $x^{-1}y \in H$. Conversely, if $x^{-1}y \in H$, then (Exercise 2, above), $(x^{-1}y)H = H$, hence $x((x^{-1}y)H) = xH$, or $yH = xH$.

$$\text{Q.E.D.}$$

Exercises

1. Let $H = \{e\}$ where e is the neutral element of G. Prove that there is a one-one correspondence between G and the left (and right) coset decomposition determined by H. (*Note:* In this case the two decompositions are the same.)

2. If $H = G$, how many equivalence classes are there in each of the two coset decompositions determined by H?

16. *Index of a Subgroup*

In this section we examine the structures of partitions determined by subgroups. Let H be a subgroup of G; in the preceding section we saw that H determines both a left and a right coset decomposition of G, namely,

$$\mathscr{P}_\ell = \{xH \,|\, x \in G\} \text{ and } \mathscr{P}_r = \{Hx \,|\, x \in G\},$$

respectively.

THEOREM 32. (i) There exists a one-one correspondence between $\mathscr{P}_\ell$ and $\mathscr{P}_r$.

(ii) For each x and each y in G, there exists a one-one correspondence between xH and yH and between Hx and Hy.

(iii) For each x and each y in G, there is a one-one correspondence between xH and Hy.

Proof: (i) First we show that $xH = yH$ if and only if $Hx^{-1} = Hy^{-1}$. Indeed, $xH = yH$ if and only if $x^{-1}y \in H$. But $x^{-1}y = x^{-1}(y^{-1})^{-1}$; hence (Theorem 31), $Hx^{-1} = Hy^{-1}$.

For each $x \in G$, define $f(xH) = Hx^{-1}$; the foregoing shows that f is a one-one mapping of $\mathscr{P}_\ell$ onto the set $\{Hx^{-1} \mid x \in G\}$. But $\mathscr{P}_r = \{Hx^{-1} \mid x \in G\}$ (why?). Therefore f is a one-one mapping of $\mathscr{P}_\ell$ onto $\mathscr{P}_r$.

(ii) It is easy to see that the mapping $g_\ell : xH \longrightarrow yH$ defined by

$$g_\ell(xh) = yh, \quad h \in H,$$

is one-one and onto, as is the mapping $g_r : Hx \longrightarrow Hy$ defined by

$$g_r(hx) = hy, \quad h \in H.$$

The details are left to the reader.

(iii) To prove (iii) observe that for each x and each y in G the mapping $t : xH \longrightarrow Hy$ defined by $t(xh) = hy$, $h \in H$, is a bijection.

Q.E.D.

COROLLARY 1.   If the order of the subgroup H is finite then all left and right cosets of H have the same number of elements.

COROLLARY 2.   If the number of elements in $\mathscr{P}_\ell$ (in $\mathscr{P}_r$) is finite, so is the number of elements in $\mathscr{P}_r$ (in $\mathscr{P}_\ell$) and $|\mathscr{P}_r| = |\mathscr{P}_\ell|$.

DEFINITION 31.   Let H be a subgroup of G such that the left coset decomposition $\mathscr{P}_\ell$ determined by H is finite. The *index of H in G* is the number of elements in $\mathscr{P}_\ell$ (in $\mathscr{P}_r$).

By Corollary 2, the index of H in G is also the number of elements in the right coset decomposition determined by H.

COROLLARY 3.   If G is a finite group and if j is the index of H in G, then $|G| = j \cdot |H|$.

Proof: Exercise.

COROLLARY 4.   (Lagrange's Theorem) If G is a finite group and if $H < G$, then $|H| \mid |G|$.

Proof: Exercise.

COROLLARY 5.   If G is a finite group and if $x \in G$, then $o(x) \mid |G|$.

Proof: Since G is a finite group, x has finite order, say, m. Therefore $K = \{x^k \,|\, k = 0, 1, \ldots, m - 1\}$ is a subgroup of G (why?) and its order is precisely m. By Corollary 4, $m\,||\,G|$, i.e., $o(x)\,||\,G|$.

Q.E.D.

17. Stable Relations, Normal Subgroups, Quotient Groups

Let G be a group and let $\sim$ be a stable equivalence relation on G. Thus $\sim$ is both left stable and right stable. Since $\sim$ is an equivalence relation on G, it determines one and only one partition $\mathscr{P}$ of G. Moreover, there is a subgroup H of G such that $H \in \mathscr{P}$ and $H = \{g \,|\, g \in G \text{ and } g \sim e\}$. Now since $\sim$ is left stable, by Theorem 31,

$$\mathscr{P} = \{xH \,|\, x \in G\};$$

on the other hand, since $\sim$ is also right stable, Theorem 31 shows that

$$\mathscr{P} = \{Hx \,|\, x \in G\}.$$

Hence

$$\{xH \,|\, x \in G\} = \{Hx \,|\, x \in G\},$$

and so for each $x \in G$, there is a $y \in G$ such that $xH = Hy$, and conversely. We claim more:

THEOREM 33.   With H as above, $xH = Hx$ for each $x \in G$.

Proof: For each x in G, the left coset of H containing x is xH; the right coset of H containing x is Hx. But since the left and right coset decompositions of H are equal, it follows that $xH = Hx$.

Q.E.D.

DEFINITION 32.   A subgroup N of G is *normal* (or *invariant*, or *self-conjugate*) if and only if for each $x \in G$,

$$xN = Nx.$$

If N is a normal subgroup of G, we write

$$N \lhd G.$$

Exercise: Let N be a subgroup of G and let $\sim_\ell$ and $\sim_r$ be the left and right stable equivalence relations, respectively, determined by N. Prove that $N \lhd G$ if and only if $\sim_\ell = \sim_r$, hence if and only if $\sim = \sim_\ell = \sim_r$ is stable. This exercise provides another criterion for a subgroup to be normal.

A further criterion for normality is given by

THEOREM 34. A subgroup N of G is a normal subgroup of G if and only if for each $x \in G$,

$$xNx^{-1} \subset N.$$

Proof: If N is a normal subgroup of G, then for each $x \in G$, $x^{-1}N = Nx^{-1}$. Hence

$$N = x(x^{-1}N) = x(Nx^{-1}) = xNx^{-1},$$

and therefore $xNx^{-1} \subset N$.

Conversely, if $xNx^{-1} \subset N$ for each $x \in G$, then $(x^{-1})N(x^{-1})^{-1} = x^{-1}Nx \subset N$. But, $N = x(x^{-1}Nx)x^{-1} \subset xNx^{-1}$. Therefore $N = xNx^{-1}$, whence $xN = Nx$ for each $x \in G$, and consequently $N \lhd G$.

Q.E.D.

Exercises

1. Prove: If H is a subgroup of an Abelian group G, then $H \lhd G$.

2. Let N be a subgroup of a group G. Prove that $N \lhd G$ if and only if for each $x \in G$ and for each $z \in N$, $xN = zxN$.

3. Let G be a group, and let $Z = \{z \mid z \in G$ and $zx = xz$ for each $x \in G\}$; Z is the *center* of G. Prove that $Z \lhd G$.

4. Let H and K be normal subgroups of G. Prove that $H \cap K \lhd G$.

5. Let $N < H$ and $H < G$. Prove: If $N \lhd G$, then $N \lhd H$.

6. Prove that the Four Group $\mathscr{F} = \{(1), (12)(34), (13)(24), (14)(23)\}$ is a normal subgroup of $\mathscr{S}_4$. Is the subgroup $H = \{(1), (12), (34), (12)(34)\}$ a normal subgroup of $\mathscr{S}_4$?

7. Prove: If H is a subgroup of G of index 2, then $H \lhd G$. (*Hint:* Show that the left and right coset decompositions of G determined by H are equal.) By Exercise 2, page 97, for $n \geq 2$, $\mathscr{A}_n \lhd \mathscr{S}_n$.

8. An element $u \in G$ is a *commutator* if and only if there exist x and y in G such that $u = xyx^{-1}y^{-1}$. One writes, "$[x, y] = xyx^{-1}y^{-1}$." Prove: If u is a commutator, then so is zuz^{-1} for each $z \in G$. (*Hint:* $z[x, y]z^{-1} = zxz^{-1}zyz^{-1}zx^{-1}z^{-1}zy^{-1}z^{-1}$.)

9. Prove that the inverse of a commutator is a commutator.

10. Let C be the set of all finite products of commutators in G. Prove that $C \lhd G$; C is the *commutator* subgroup of G. Note that C is the subgroup of G generated by the set of all commutators.

11. Compute the center and the commutator subgroup of $\mathscr{S}_3$; of $\mathscr{S}_4$.

If N is a normal subgroup of G, then N determines left and right coset decompositions of G, namely, $\{xN \mid x \in G\}$ and $\{Nx \mid x \in G\}$, respectively. But since $xN = Nx$ for each $x \in G$, $\{xN \mid x \in G\} = \{Nx \mid x \in G\}$.

DEFINITION 33. If N is a normal subgroup of G, the partition $\{xN \mid x \in G\}$ is the *coset decomposition of G determined by N*; it is denoted by "G/N,"

$$G/N = \{xN \mid x \in G\} \quad (= \{Nx \mid x \in G\}.)$$

The elements in G/N are the *cosets* of N.

Using the coset decomposition G/N, we shall construct a new group—related to G in important ways—by defining a binary operation on G/N. As a preliminary step, we require

LEMMA 1. Let N be a subgroup of G. Then N is a normal subgroup of G if and only if the following condition holds:

(†) for all x, y, x', y' in G, if $xN = x'N$ and $yN = y'N$, then $xyN = x'y'N$.

> *Proof:* Suppose, first, $N \lhd G$. By Theorem 31, it suffices to prove that $(xy)^{-1}(x'y') \in N$. Now $(xy)^{-1}(x'y') = y^{-1}(x^{-1}x')y'$, and since $xN = x'N$, $x^{-1}x' \in N$. Hence $y^{-1}(x^{-1}x')y' \in y^{-1}Ny'$. But since $N \lhd G$, $Ny' = y'N$; and further, since $yN = y'N$, $y^{-1}Ny' = y^{-1}(yN) = N$. Therefore $(xy)^{-1}(x'y') \in N$.
>
> Conversely, suppose the condition (†) holds. Let $z \in N$ and let $x \in G$. Then $N = eN = zN$ and $xN = xN$. Hence $xN = exN = zxN$ and by Exercise 2, page 112, $N \lhd G$.
>
> Q.E.D.

DEFINITION 34. Let N be a normal subgroup of G, and let $A, B \in G/N$. If $a \in A$, $b \in B$, let C be the coset of N containing ab. Then set

$$A * B = C.[4]$$

C is the *product* of the cosets A and B.

LEMMA 2. If $A, B \in G/N$ and if $a \in A$, $b \in B$, then

$$A * B = (aN) * (bN) = (ab)N.$$

> *Proof:* Exercise.

LEMMA 3. $*$ is a binary operation on G/N.

> *Proof:* Definition 34 assures us that if $A, B \in G/N$, then $A * B \in G/N$. Moreover, if $A = A'$, $B = B'$, then $A = aN$, $A' = a'N$, $B = bN$, $B' = b'N$ where $a, a' \in A = A'$ and $b, b' \in B = B'$. By Lemma 1
>
> $$A * B = (ab)N = (a'b')N = A' * B',$$
>
> and so $*$ is a binary operation on G/N.
>
> Q.E.D.

---

[4] A simpler and more elegant definition of product of cosets is $A * B = \{xy \mid x \in A$ and $y \in B\}$. However, this alternative definition is not appropriate in the study of other algebraic structures such as rings and vector spaces, whereas this one can be carried over.

THEOREM 35. Let N be a normal subgroup of G. Then the structure consisting of G/N and the binary operation $*$ is a group.

Proof: In view of Lemma 3, all we need do is to check that conditions G.1—G.3 of Definition 1 are satisfied.

If $aN, bN, cN \in G/N$, then

$$(aN * bN) * cN = (ab)N * cN$$
$$= (ab)cN$$
$$= a(bc)N$$
$$= aN * (bc)N$$
$$= aN * (bN * cN),$$

hence G.1 holds. Clearly, N itself is a left neutral element for G/N and this verifies G.2. Finally, for each $a \in G$, $a^{-1}N * aN = (a^{-1}a)N = eN = N$ so that each element in G/N has a left inverse. This establishes G.3 and completes the proof that G/N together with the binary operation $*$ constitutes a group.

Q.E.D.

DEFINITION 35. If N is a normal subgroup of G, then the group consisting of G/N and the binary operation $*$ is the *quotient* (or, *factor*) group of G with respect to N. If G is an additive group, one frequently uses the term "difference group" instead of "factor group" and the symbol "$G - N$" in place of "G/N"; also in this case we replace the notation "$*$" for the binary operation on G/N by "$\oplus$."

EXAMPLE 22. Let $G = \mathscr{S}_3$ and let $\mathscr{A}_3$ be the alternating group, $\mathscr{A}_3 = \{(1), (123), (132)\}$; $\mathscr{A}_3$ is a subgroup of $\mathscr{S}_3$ of index 2, therefore $\mathscr{A}_3 \lhd \mathscr{S}_3$. The cosets are $\mathscr{A}_3$ and $(12)\mathscr{A}_3 = \{(12), (23), (31)\}$. Thus the quotient group $\mathscr{S}_3/\mathscr{A}_3$ is the group whose set of elements is $\{\mathscr{A}_3, (12)\mathscr{A}_3\}$ and the binary operation $*$ on $\mathscr{S}_3/\mathscr{A}_3$ is given by

$$\mathscr{A}_3 * \mathscr{A}_3 = \mathscr{A}_3\mathscr{A}_3 = \mathscr{A}_3,$$
$$\mathscr{A}_3 * (12)\mathscr{A}_3 = \mathscr{A}_3((12)\mathscr{A}_3) = (12)\mathscr{A}_3$$
$$= (12)\mathscr{A}_3 * \mathscr{A}_3,$$
$$(12)\mathscr{A}_3 * (12)\mathscr{A}_3 = \mathscr{A}_3.$$

EXAMPLE 23. Let $G = \mathscr{S}_n$ and let $\mathscr{A}_n$ be the alternating group. Since (12) is an odd permutation, $(12) \notin \mathscr{A}_n$, and since $\mathscr{A}_n$ is of index 2, the coset decomposition determined by $\mathscr{A}_n$ is

$$\{\mathscr{A}_n, (12)\mathscr{A}_n\},$$

and this is the set of the quotient group $\mathscr{S}_n/\mathscr{A}_n$. With the group operation $*$, we have $\mathscr{A}_n * \mathscr{A}_n = \mathscr{A}_n$, $\mathscr{A}_n * (12)\mathscr{A}_n = (12)\mathscr{A}_n = (12)\mathscr{A}_n * \mathscr{A}_n$ and $(12)\mathscr{A}_n * (12)\mathscr{A}_n = \mathscr{A}_n$.

EXAMPLE 24.   Let Q be the additive group of rational numbers. The additive group of integers Z is a subgroup of Q and since Q is commutative, $Z \lhd Q$. The quotient group Q/Z consists of the set

$$\{x + Z \mid x \in Q\},$$

and the binary operation $\oplus$ where

$$(x + Z) \oplus (y + Z) = (x + y) + Z.$$

We claim that the group Q/Z has infinite order, but that each element in Q/Z has finite order. Indeed,

$$x + Z = y + Z$$

if and only if $(-y) + x = x - y \in Z$, i.e., if and only if the rational numbers x and y differ by an integer. Hence if $m \neq n$ are positive integers, the cosets $1/m + Z$ and $1/n + Z$ are distinct. Thus

$$\left\{\frac{1}{k} + Z \mid k \in Z\right\}$$

is an infinite subset of QZ, and so Q/Z has infinite order.

To show that each element in Q/Z has finite order, let $x + Z \in Q/Z$ where $x = p/q$ and $p, q \in Z$, $q \neq 0$. Then

$$q\left(\frac{p}{q} + Z\right) = \underbrace{\left(\frac{p}{q} + Z\right) * \cdots * \left(\frac{p}{q} + Z\right)}_{q \text{ times}}$$

$$= p + Z$$

$$= Z;$$

since Z is the neutral element of Q/Z, $x + Z$ has finite order.

Exercises

1. Let G be the group of the square (Exercises 6 and 7, pages 56 and 57); show that $N = \{I, R, S, T\}$ is a normal subgroup of G and describe G/N.

2. Let G be the group defined in Exercise 4, page 56, and let $H = \{(a, 0) \mid a \in Z\}$. Show that $H \lhd G$ and describe G/H.

3. Let N be a normal subgroup of G. Prove: If $\bar{x}, \bar{y} \in G/N$, then $\bar{x} * \bar{y} = \overline{xy}$ where for each $z \in G$, $\bar{z} = zN$ is the coset of N containing z.

4. Let $G_1, G_2, \ldots, G_n, n \geq 1$, be n (multiplicative) groups with neutral elements $e_i, 1 \leq i \leq n$, respectively, and define $*: (G_1 \times G_2 \times \cdots \times G_n) \times (G_1 \times G_2 \times \cdots \times G_n) \longrightarrow G_1 \times G_2 \times \cdots \times G_n$ by

$$(a_1, a_2, \ldots, a_n) * (b_1, b_2, \ldots, b_n) = (a_1 b_1, a_2 b_2, \ldots, a_n b_n),$$

where $a_i, b_i \in G_i$, $1 \leq i \leq n$.

  a) Prove that $G = G_1 \times G_2 \times \cdots \times G_n$ is a group with group operation $*$.
  b) Set $\bar{G}_i = \{(e_1, \ldots, e_{i-1}, a_i, e_{i+1}, \ldots, e_n) \mid a_i \in G_i\}$, $1 \leq i \leq n$. Prove that each $\bar{G}_i$ is a normal subgroup of G.

c) Describe each $G/\bar{G}_i$, $1 \leq i \leq n$.

5. Prove: If G is an Abelian group and $H < G$, then G/H is Abelian.

6. In elementary theory of equations, it is shown that the n roots of the polynomial equation $x^n - 1 = 0$, $(n > 0)$, are given by

$$\zeta_k = \cos\frac{2k\pi}{n} + i\sin\frac{2k\pi}{n}, \quad 0 \leq k \leq n - 1.$$

These are the *n-th roots of unity*. By elementary trigonometry we have:

(25) $(\cos\alpha + i\sin\alpha)(\cos\beta + i\sin\beta) = \cos(\alpha + \beta) + i\sin(\alpha + \beta)$.

Using the multiplication (25) as group operation, prove that the n-th roots of unity form a group.

We conclude this section with a useful application of the concept of the quotient group to the theory of the symmetric groups.

THEOREM 36.   If $n \geq 2$, then the only subgroup of index 2 in $\mathscr{S}_n$ is the alternating group $\mathscr{A}_n$.

LEMMA.   If H is a subgroup of G of index 2, then for each $x \in G$, $x^2 \in H$.

Proof: If $x \in H$, then obviously $x^2 \in H$.

Suppose $x \notin H$. Since the index of H in G is 2, $H \vartriangleleft G$ (Exercise 7, page 112) and the quotient group G/H is a group consisting of two elements, $\bar{e} = H$ and $\bar{x} = xH$. Thus $\bar{x}^2 = \bar{e}$, i.e., $\bar{x}^2 = (xH)^2 = x^2H = H$, whence $x^2 \in H$.

Proof of Theorem 36. Suppose H is a subgroup of $\mathscr{S}_n$ of index 2. If $n = 2$, it is clear that $H = \mathscr{A}_2$.

Henceforth let $n \geq 3$ and assume we have proved that every 3-cycle $(12k)$, $3 \leq k \leq n$, is in H. Since the set of 3-cycles $\{(12k) \mid 3 \leq k \leq n\}$ generates $\mathscr{A}_n$, it follows that $\mathscr{A}_n \subset H$. On the other hand, since $\mathscr{A}_n$ and H have the same index in $\mathscr{S}_n$, they also have the same order, whence $\mathscr{A}_n = H$. All that's left is to prove that each of the 3-cycles $(12k)$, $3 \leq k \leq n$, is in H. But $(12k) = (1k2)^2$; hence by the lemma each $(12k) \in H$ and the theorem is proved.

Q.E.D.

18.   *Conclusion*

In Chapter 2 we introduced some of the basic concepts of elementary group theory and illustrated these concepts with examples and exercises. The view presented here is an internal one; we have been concerned with group elements and subgroups of groups. An external view of a group is obtained by making comparisons of the given group with other groups. The tools for such investigations are known as "isomorphisms" and "homomorphisms"; these will be the subjects of Chapter 3. It will be seen that the internal structure of a group is related to the external comparisons, and that the normal subgroups and factor groups are basic for the study of homomorphisms.

3

Group Isomorphism
and Homomorphism

1. *Introduction*

In the preceding chapter we initiated the study of groups. The view taken of these structures was an internal one. Given a group, what could one say of its elements, its subgroups, its normal subgroups, and so on. In the present chapter we extend our knowledge of groups by making comparisons among various structures, and in this sense we may consider the study as external to a given group. The two views, internal and external, are by no means divorced from each other. On the contrary, it will turn out that internal structure and external comparisons are closely related.

The standard tools for making comparisons among groups are mappings known as "homomorphisms" and "isomorphisms." Isomorphism is a special case of homomorphism. Thus a completely orderly study of these mappings would begin with an investigation of homomorphisms in general, and at the proper place isomorphism would be defined and many of its properties deduced from the available knowledge of homomorphisms. On the other hand, from the learner's point of view, the concept of isomorphism is perhaps a little more immediate than that of homomorphism. For this reason, we shall alter the logical order of things and begin with group isomorphisms.

2. Group Isomorphism; Examples, Definitions, and Simplest Properties.

EXAMPLE 1. As usual, let $\mathscr{S}_3$ be the group of all permutations of the set $\{1, 2, 3\}$ and let M be the group of rigid motions of the triangle (Figure 6). Thus (Example 3, Chapter 2) $M = \{I, R, T, F_a, F_b, F_c\}$ and the group operation $*$ of the group M is completely defined by Table 6.

TABLE 6

$*$	I	R	T	F_a	F_b	F_c
I	I	R	T	F_a	F_b	F_c
R	R	T	I	F_c	F_a	F_b
T	T	I	R	F_b	F_c	F_a
F_a	F_a	F_b	F_c	I	R	T
F_b	F_b	F_c	F_a	T	I	R
F_c	F_c	F_a	F_b	R	T	I

Since the groups $\mathscr{S}_3$ and M have the same order, there is evidently a bijection $M \longrightarrow \mathscr{S}_3$; indeed, there are precisely six such one-one correspondences. Among these six bijections there is one which is particularly noteworthy. Namely, with each motion X in the group M, we associate exactly one permutation ξ in $\mathscr{S}_3$ in the following way:

(1)
$$\begin{cases} I & \longleftrightarrow (1), \\ R & \longleftrightarrow (123), \\ T & \longleftrightarrow (132), \\ F_a & \longleftrightarrow (23), \\ F_b & \longleftrightarrow (13), \\ F_c & \longleftrightarrow (12). \end{cases}$$

This association defines a one-one correspondence between the two groups under consideration, and the correspondence thus defined has a striking feature. Namely, *if $X \longleftrightarrow \eta$ and if $Y \longleftrightarrow \eta$, then $X * Y \longleftrightarrow \xi\eta$.* For instance,

$$R \longleftrightarrow (123),$$
$$F_a \longleftrightarrow (23),$$
$$R * F_a = F_c,$$
$$(123)(23) = (12),$$

and by (1),

$$F_c \longleftrightarrow (12);$$

in other terms,

$$R * F_a \longleftrightarrow (123)(23).$$

Similarly,

$$F_c \longleftrightarrow (12),$$
$$T \longleftrightarrow (132),$$
$$F_c * T = F_b,$$
$$(12)(132) = (13),$$

and by (1),

$$F_c * T = F_b \longleftrightarrow (13) = (12)(132).$$

The reader is urged to verify: For each X and for each Y, if $X \longleftrightarrow \xi$ and $Y \longleftrightarrow \eta$, then $X * Y \longleftrightarrow \xi\eta$.

Now, it can hardly be maintained that a motion of a piece of cardboard and a mapping of some particular set are the same thing. Yet the groups M (consisting of certain motions of a triangular cardboard) and $\mathscr{S}_3$ (consisting of certain mappings of the set $\{1, 2, 3\}$) are alike in very important ways. Not only do the two groups have the same order but, in addition, if we use the correspondence defined by (1) to replace each $X \in M$ by its associate $\xi \in \mathscr{S}_3$ in the group table for M, then the resulting table is the group table for $\mathscr{S}_3$. Thus M and $\mathscr{S}_3$ "behave" in the same way relative to the respective group operations; one says that the given bijection "preserves the group operation." If one wishes to study the motions of a triangle, then $\mathscr{S}_3$ provides a simplified model of M from which much extraneous detail is excluded. Conversely, M provides a geometrical interpretation for $\mathscr{S}_3$.

Exercise: Exhibit several bijections $M \longrightarrow \mathscr{S}_3$ which do not preserve the group operation.

EXAMPLE 2. The mere fact that two groups have the same order is not sufficient to insure that they "behave" in the same way relative to their respective group operations. For instance, let $G = Z_2 \times Z_2$ and let the group operation $*$ be defined by

$$(a, b) * (c, d) = (a \oplus c, b \oplus d),$$

where (a, b) and $(c, d) \in G$. The order of G is 4. The modular arithmetic Z_4 is a group of order 4 with respect to $\oplus$ as group operation, and so $|G| = |Z_4| = 4$. There are precisely $4! = 24$ one-one mappings between G and Z_4, and no matter which of these correspondences is selected there will always be an X and a Y in G, an $\bar{a}$ and a $\bar{b}$ in Z_4 such that

$$X \longleftrightarrow \bar{a} \quad \text{and} \quad Y \longleftrightarrow \bar{b},$$

but that

$$X * Y \not\longleftrightarrow \bar{a} \oplus \bar{b}.$$

At the present stage, the verification of this statement entails the monumental job of examining all 24 one-one mappings. In a little while, we shall see that the result follows from some elementary theoretical considerations. (See Exercise 4, page 123.)

The one-one correspondence defined by (1) is an example of a *group isomorphism*.

DEFINITION 1. Let G be a (multiplicative) group, and let H be a set with binary operation $*$. An injection $\sigma: G \longrightarrow H$ is a *monomorphism* if and only if for each x and each y in G,

(2) $$\sigma(xy) = \sigma(x) * \sigma(y).$$

As usual, let $\sigma[G] = \{\sigma(x) \,|\, x \in G\}$; then $\sigma[G] \subset H$. If $\sigma[G] = H$, then σ is an *isomorphism* between G and H. In any case, the set $\sigma[G]$ is *the isomorphic image of G under σ*. Equation (2) expresses the *morphism* property of σ.

Obviously, every isomorphism is a monomorphism; the converse, of course, is not true.

In case G is an additive group, equation (2) becomes

(2′) $$\sigma(x + y) = \sigma(x) * \sigma(y).$$

In terms of Definition 1, we now see that the mapping defined by (1) (page 119) is an isomorphism. In the exercise of page 120, one shows that not every bijection $M \longleftrightarrow \mathscr{S}_3$ is an isomorphism. And in Example 2, page 120, it is asserted (without proof) that of the 24 possible bijections between the groups under consideration not even one is an isomorphism, hence the groups are not isomorphic.

Definition 1 leaves an obvious question unanswered: "What can be said about the set H and the binary operation $*$?"

THEOREM 1. Let G be a group with neutral element e, let H be a set with binary operation $*$ and let $\sigma: G \longrightarrow H$ be a monomorphism. Further, let $\circledast$ be the restriction of $*$ to $\sigma[G]$. Then $\sigma[G]$ is a group with group operation $\circledast$. The neutral element of $\sigma[G]$ is $\sigma(e)$, and for each element $\sigma(x) \in \sigma[G]$ the inverse is $\sigma(x^{-1})$, i.e., for each $x \in G$, $\sigma(x^{-1}) = (\sigma(x))^{-1}$.

The statement of Theorem 1 is usually abbreviated by the more compact sentence:

An isomorphic image of a group is a group.

Proof: The first step is to prove that $\circledast$ is a binary operation on $\sigma[G]$. By definition of restriction, if $u, v \in \sigma[G]$, then

$$u \circledast v = u * v.$$

Hence to prove that $\circledast$ is a binary operation on $\sigma[G]$, it suffices to show that

$$u * v \in \sigma[G],$$

for all $u, v \in \sigma[G]$. Since $u, v \in \sigma[G]$, there are elements $x, y \in G$ such that $\sigma(x) = u$ and $\sigma(y) = v$. Since G is a group, $xy \in G$, and so $\sigma(xy) \in \sigma[G]$. Then

$$u * v = \sigma(x) * \sigma(y)$$
$$= \sigma(xy),$$

by Definition 1, and therefore $u * v \in \sigma[G]$. This shows that $\circledast$ is a binary operation on G.

Next, $\circledast$ is associative. For, if $u, v, w \in \sigma[G]$, then there are elements $x, y, z \in G$ such that

$$\sigma(x) = u, \ \sigma(y) = v, \ \sigma(z) = w.$$

Then

$$u \circledast (v \circledast w) = \sigma(x) \circledast (\sigma(y) \circledast \sigma(z))$$
$$= \sigma(x) \circledast \sigma(yz)$$
$$= \sigma(x(yz))$$
$$= \sigma((xy)z)$$
$$= \sigma(xy) \circledast \sigma(z)$$
$$= (\sigma(x) \circledast \sigma(y)) \circledast \sigma(z)$$
$$= (u \circledast v) \circledast w.$$

Further, $\sigma(e)$ is a left neutral element for $\sigma[G]$. Indeed, if $u \in \sigma[G]$, then there is an $x \in G$ such that $\sigma(x) = u$ and

$$\sigma(e) \circledast u = \sigma(e) \circledast \sigma(x)$$
$$= \sigma(ex)$$
$$= \sigma(x)$$
$$= u.$$

Finally, each $u \in \sigma[G]$ has a left inverse in $\sigma[G]$. For, if $\sigma(x) = u$ where $x \in G$, then

$$\sigma(e) = \sigma(x^{-1}x)$$
$$= \sigma(x^{-1}) \circledast \sigma(x)$$
$$= \sigma(x^{-1}) \circledast u,$$

and so $\sigma(x^{-1})$ is a left inverse of $u = \sigma(x)$, i.e., $\sigma(x^{-1}) = (\sigma(x))^{-1}$.

Q.E.D.

DEFINITION 2. Let G and H be groups; G *is isomorphic to H* if and only if there exists an isomorphism between G and H. If G is isomorphic to H, one writes:

$$G \cong H.$$

Exercises

1. Let $G = \{1, -1, i, -i\}$ with ordinary multiplication of complex numbers as group operation. Prove that G is isomorphic to Z_4 (with group operation $\circledast$).

2. Find an isomorphism, other than (1), between M (see Example 1) and $\mathscr{S}_3$. This exercise shows that if G is isomorphic to H, then there may be more than one isomorphism between G and H.

3. Prove that an isomorphic image of an Abelian group is Abelian.

4. Prove that an isomorphic image of a cyclic group is cyclic. Hence, show that the group G (Example 2, page 120) is not isomorphic to the modular arithmetic Z_4.

5. Prove that cyclic groups of the same order are isomorphic. From this exercise, it follows that every cyclic group of order n is isomorphic to the modular arithmetic Z_n, and that every infinite cyclic group is isomorphic to the additive group of the integers.

6. Prove that isomorphism is reflexive, symmetric and transitive. Hence isomorphism is an equivalence relation on every set of groups.

7. Let a be an element in a group G and define $\sigma_a : G \longrightarrow G$ by

$$\sigma_a(x) = axa^{-1}$$

for each $x \in G$. Prove that σ_a is an isomorphism.

8. Examine the proof of Theorem 1 very carefully to determine whether or not all the properties of isomorphism were used. Suggest (and prove!) a generalization of Theorem 1.

9. Let $H < G$ and let f be an isomorphism of G. Prove that $f[H] < f[G]$. If the index of H in G is finite, show that this index is equal to the index of $f[H]$ in $f[G]$.

10. Let f be an isomorphism of a group G, and let H and K be distinct subgroups of G. Prove that $f[H]$ and $f[K]$ are distinct subgroups of $f[G]$. Further, show that if $N \triangleleft G$, then $f[N] \triangleleft f[G]$.

Remark: It is interesting to observe that $H = \{(1), (12), (34), (12)(34)\}$ and $\mathscr{F}_4 = \{(1), (12)(34), (13)(24), (14)(23)\}$ are isomorphic subgroups of $\mathscr{S}_4$. Yet although $\mathscr{F}_4$ is normal in $\mathscr{S}_4$, H is not a normal subgroup of $\mathscr{S}_4$. Does this fact contradict the result proved in Exercise 10?

11. Let S be a set of generators of G, and let f be an isomorphism of G. Prove that $f[S]$ is a set of generators of $f[G]$.

In view of the fact that isomorphism is symmetric (Exercise 6, above) it is appropriate to introduce

DEFINITION 2′. Groups G and H are *isomorphic to each other* (or, simply, *isomorphic*) if and only if G is isomorphic to H.

EXAMPLE 3. An isomorphism enjoying great popularity is provided by the natural logarithm. This is a novel but useful and interesting way of looking at logs, so let's try it. Let R_+ be the additive group of the real numbers, and let R'_+ be the multiplicative group of the positive real numbers. Further, let

$f: R'_+ \longrightarrow R_+$ be the mapping

$$f(x) = \log_e x, \quad x \in R'_+;$$

since every positive real number possesses a natural logarithm,[1] the mapping f is defined for each $x \in R'_+$. Also, since $\log_e x = \log_e y$ if and only if $x = y$, f is one-one. Moreover, if $z \in R_+$, then $f(e^z) = \log_e e^z = z$ so that f is onto. Finally, if $x, y \in R'_+$, then

$$\begin{aligned} f(xy) &= \log_e(xy) \\ &= \log_e x + \log_e y \\ &= f(x) + f(y). \end{aligned}$$

Hence f is an isomorphism of the multiplicative group R'_+ onto the additive group R_+.

The mapping f defined in this example is only one of many isomorphisms between the multiplicative group R'_+ of positive reals and the additive group of the real numbers. Indeed, for each real number b, $b > 1$, the mapping

$$f_b(x) = \log_b x, \quad x \in R'_+,$$

is again an isomorphism between the two given groups. Note that every one of the isomorphisms f_b has the property that $f_b(1) = \log_b 1 = 0$ (why?). If we take $b = e$, then $f_e = f$. This example shows that if two groups are isomorphic, then there may be more than one isomorphism between them.

3. The Isomorphism Theorem for the Symmetric Groups

We now apply the foregoing ideas to answer a question raised in Chapter 2; namely, if $\mathscr{S}_n$ is the group of all permutations of the n-element set $A_n = \{x_1, x_2, \ldots, x_n\}$ and $\mathscr{S}'_n$ is the group of all permutations of the n-element set $B_n = \{y_1, y_2, \ldots, y_n\}$, what can one say about relations between $\mathscr{S}_n$ and $\mathscr{S}'_n$? The answer is

THEOREM 2. $\mathscr{S}_n$ and $\mathscr{S}'_n$ are isomorphic.

Proof: We prove that there exists a mapping $f: \mathscr{S}_n \longrightarrow \mathscr{S}'_n$ which is, in fact, an isomorphism.

Since A_n and B_n have the same number of elements, there exists a bijection $\varphi: A_n \longrightarrow B_n$. Now for each $\sigma \in \mathscr{S}_n$ set

$$f(\sigma) = \varphi \circ \sigma \circ \varphi^{-1};$$

we claim that f is an isomorphism between $\mathscr{S}_n$ and $\mathscr{S}'_n$.

First, $f(\sigma) \in \mathscr{S}'_n$. For, since φ and σ are one-one mappings,

---

[1] Several deep results have been hidden in the assertions made here concerning the mapping f. For a detailed discussion of these matters see *Calculus*, T. M. Apostol, Vol. 1, Boston: Blaisdell, 1961, pp. 174–182.

so is $\varphi \circ \sigma \circ \varphi^{-1}$. And if $y \in B_n$, then $(\varphi \circ \sigma \circ \varphi^{-1})(y) \in B_n$ (details?). Hence for each $\sigma \in \mathscr{S}_n$, $f(\sigma) \in \mathscr{S}_n'$. Next, if σ, τ are permutations in $\mathscr{S}_n$, a simple computation shows that $f(\sigma) = f(\tau)$ if and only if $\sigma = \tau$, and so f is injective. Further, if $\sigma' \in \mathscr{S}_n'$, then $\varphi^{-1} \circ \sigma' \circ \varphi \in \mathscr{S}_n$ (why?) and clearly $f(\varphi^{-1} \circ \sigma' \circ \varphi) = \sigma'$. This shows that f is onto.

Finally, f is an isomorphism. Indeed, if $\sigma, \tau \in \mathscr{S}_n$, then

$$f(\sigma\tau) = \varphi \circ (\sigma\tau) \circ \varphi^{-1}$$
$$= (\varphi \circ \sigma \circ \varphi^{-1}) \circ (\varphi \circ \tau \circ \varphi^{-1})$$
$$= f(\sigma) f(\tau),$$

and so f satisfies the condition of Definition 1.

Q.E.D.

Suppose $\mathscr{A}_n$ and $\mathscr{A}_n'$ are the alternating groups in $\mathscr{S}_n$ and $\mathscr{S}_n'$, respectively. Then (Theorem 36, Chapter 2) $\mathscr{A}_n$ is the only subgroup of index 2 in $\mathscr{S}_n$, and $\mathscr{A}_n'$ is the only subgroup of index 2 in $\mathscr{S}_n'$. Now if f is an isomorphism between $\mathscr{S}_n$ and $\mathscr{S}_n'$, then (Exercise 9, page 123) $f[\mathscr{A}_n]$ has index 2 in $\mathscr{S}_n'$, hence $f[\mathscr{A}_n] = \mathscr{A}_n'$. Thus we have the

COROLLARY. $\mathscr{A}_n \cong \mathscr{A}_n'$.

Usually, as a matter of convenience in notation, $\mathscr{S}_n$ is regarded as the symmetric group acting on $I_n = \{1, 2, \ldots, n\}$. In this case, the elements of $\mathscr{S}_n$ are denoted by

$$(12), (1234), \begin{pmatrix} 123 \ldots n \\ 342 \ldots \end{pmatrix}, \text{ etc.}$$

The set $\{(1k) \mid 2 \leq k \leq n\}$ is a set of generators of $\mathscr{S}_n$ (see Exercise 3, page 82), and the set $\{(12k) \mid 3 \leq k \leq n\}$ is a set of generators of $\mathscr{A}_n$ (Exercise 5, page 97).

Theorem 2 can be generalized in the following way:

THEOREM 3. Let A and B be sets having the same cardinality. Then the full transformation groups $\mathscr{P}(A)$ and $\mathscr{P}(B)$ are isomorphic.

The proof of Theorem 3 is essentially the same as that of Theorem 2 and is left to the reader.

Exercises

1. Let $\lambda_{(123)} \colon \mathscr{S}_3 \longrightarrow \mathscr{S}_3$ be defined by

$$\lambda_{(123)}(\beta) = (123)\beta, \quad \beta \in \mathscr{S}_3.$$

Thus, in case

$$\beta = (1), \lambda_{(123)}((1)) = (123)(1) = (123),$$
$$\beta = (12), \lambda_{(123)}((12)) = (123)(12) = (13),$$
$$\beta = (23), \lambda_{(123)}((23)) = (123)(23) = (12),$$

etc. Prove that $\lambda_{(123)}$ is a bijection, hence $\lambda_{(123)} \in \mathscr{P}(\mathscr{S}_3)$. Is $\lambda_{(123)}$ an isomorphism? Why?

2. For each $\alpha \in \mathscr{S}_3$, define $\lambda_\alpha : \mathscr{S}_3 \longrightarrow \mathscr{S}_3$ by

$$\lambda_\alpha(\beta) = \alpha\beta, \quad \beta \in \mathscr{S}_3.$$

Prove that $\lambda_\alpha \in \mathscr{P}(\mathscr{S}_3)$.

3. Let G be a (multiplicative) group, and for each $a \in G$ define $\lambda_a : G \longrightarrow G$ by

$$\lambda_a(x) = ax, \ x \in G,$$

and $\rho_a : G \longrightarrow G$ by

$$\rho_a(x) = xa, \ x \in G.$$

Prove that λ_a and ρ_a are both elements in $\mathscr{P}(G)$.

4. *The Theorem of Cayley*

The study of the symmetric and the full transformation groups was introduced with the assertion that "the symmetric and the full transformation groups are rich sources of examples of groups of all kinds." The following theorem shows the remarkable extent to which this is true.

THEOREM 4. (Cayley). Each group is isomorphic with a subgroup of some full transformation group.

Note that there is no restriction on the order of the given group G. If G is finite, say $|G| = n$, the theorem will show that G is isomorphic with a subgroup of $\mathscr{S}_n$.

Proof: Given a group G, our object is to find a set A such that $G \cong H$ where H is a subgroup of $\mathscr{P}(A)$. The main idea of the proof is to take as the set A the set of elements in the group G itself! Thus we shall prove that G is isomorphic to a subgroup of the full transformation group $\mathscr{P}(G)$.

With each $a \in G$ we associate, in a "natural" way, a mapping $\lambda_a : G \longrightarrow G$, defined thus (see Exercise 3, above):

$$\lambda_a(x) = ax, \quad x \in G.$$

Clearly, $\lambda_a(x) = \lambda_a(y)$ if and only if $x = y$, hence each λ_a is an injection. Moreover, λ_a is a surjection. Indeed, if $z \in G$, then $a^{-1}z \in G$ and so

$$\lambda_a(a^{-1}z) = a(a^{-1}z) = (aa^{-1})z = ez = z.$$

Thus for each $a \in G$, λ_a is an element in $\mathscr{P}(G)$.

Now let

$$H = \{\lambda_a \mid a \in G\};$$

the above reasoning shows that H is a nonempty subset of $\mathscr{P}(G)$. The next step is to exhibit an isomorphism $f \colon G \longrightarrow H$, where the binary operation on H is composition of mappings.

For each $a \in G$ set

$$f(a) = \lambda_a;$$

we claim that f is an isomorphism. First, $f(a) = f(b)$ if and only if $a = b$. For, since λ_a and λ_b are in $\mathscr{P}(G)$, they are equal if and only if

$$\lambda_a(x) = \lambda_b(x)$$

for each $x \in G$, i.e., if and only if

$$ax = bx, \quad x \in G.$$

Since a, b and x are in G, it follows that $ax = bx$ if and only if $a = b$. Hence $f(a) = f(b)$ if and only if $\lambda_a = \lambda_b$, and so f is an injection. Further, the definitions of H and of f show that f is a surjection. Finally, if $a, b \in G$, then $f(ab) = \lambda_{ab}$ where

$$\lambda_{ab}(x) = (ab)x, \quad x \in G.$$

But

$$
\begin{aligned}
(ab)x &= a(bx) \\
&= a(\lambda_b(x)) \\
&= \lambda_a(\lambda_b(x)) \\
&= (\lambda_a \circ \lambda_b)(x),
\end{aligned}
$$

hence $\lambda_{ab} = \lambda_a \circ \lambda_b$, i.e., $f(ab) = f(a) \circ f(b)$. Hence f is an isomorphism between G and H.

By Theorem 1, H is a group with respect to composition of mappings and therefore H is a subgroup of $\mathscr{P}(G)$.

Q.E.D.

If G is a finite group, say $|G| = n$, then $\mathscr{P}(G)$ is isomorphic with $\mathscr{S}_n$ and therefore H is isomorphic with a subgroup of $\mathscr{S}_n$. Since isomorphism is transitive, it follows that G is isomorphic with a subgroup of $\mathscr{S}_n$.

Exercises

1. Let G be a four-element group with neutral element e, and let $S = \{a, b\}$ be a two-element set of generators of G. If a and b satisfy $a^2 = b^2 = e$ and $ab = ba$, exhibit an isomorphism between G and a subgroup of some $\mathscr{S}_n$.

2. Let G be a nonempty set with binary operation $*$ and let $\lambda_a \colon G \longrightarrow G$ be defined by $\lambda_a(x) = a * x$, $x \in G$. Further, let $H = \{\lambda_a \mid a \in G\}$. Prove: G is a group with respect to $*$ if and only if H is a group with respect to composition of mappings such that $\lambda_a \circ \lambda_b = \lambda_{ab}$ for each a and each b in G.

5. *Group Homomorphisms*

As Exercise 8, page 123, hints broadly, not all the properties of isomorphism are utilized in the proof of Theorem 1. Indeed, a detailed examination of the proof reveals that the one-oneness of σ is not used. The proof remains valid as it stands if one merely assumes σ to be a mapping such that the morphism property, equation (2), holds.

DEFINITION 3.   Let G be a group, and let H be a set with binary operation $*$. A mapping $\eta: G \longrightarrow H$ is a *homomorphism* if and only if for all x and y in G,

(3) $\eta(xy) = \eta(x) * \eta(y).$

Equation (3) expresses the *morphism* property of η. The homomorphism η is an *epimorphism* if and only if $\eta[G] = H$. In either case, $\eta[G]$ is the *epimorphic image* of G under η. If $A \subset H$, then $\eta^{-1}[A] = \{x \mid x \in G \text{ and } \eta(x) \in A\}$ is the *inverse-image* (or, *counter-image*) of A.

Without further ado we have

THEOREM 5.   Let G be a group with neutral element e, let H be a set with binary operation $*$ and let $\eta: G \longrightarrow H$ be a homomorphism. Then $\eta[G]$ is a group with neutral element $\eta(e)$. If $\eta(x) \in \eta[G]$, then its inverse is $\eta(x^{-1})$, i.e., $(\eta(x))^{-1} = \eta(x^{-1})$.

In short: An epimorphic image of a group is a group.

Proof: Throughout the proof of Theorem 1, replace "one-one correspondence" by "surjection." The result is a valid proof of Theorem 5.

                                                                Q.E.D.

EXAMPLE 4.   Let G and H be groups, and let e be the neutral element of H. Define $\eta: G \longrightarrow H$ by

$$\eta(x) = e,$$

for each $x \in G$. Then

$$\eta(xy) = e = e * e = \eta(x) * \eta(y),$$

hence η is a homomorphism; it is a *trivial* homomorphism.

EXAMPLE 5.   Let G be the additive group Z of the integers, and let H be the modular arithmetic Z_n, $n > 0$. We shall exhibit an epimorphism $\nu_n: Z \longrightarrow Z_n$. For each $x \in Z$, there is a unique coset of nZ containing the integer x, namely, $\bar{x} = x + nZ$. Define $\nu_n: Z \longrightarrow Z_n$ by

$$\nu_n(x) = \bar{x},$$

for each $x \in G$. For instance, if $n = 3$, then

$$\nu_3(0) = \bar{0}, \quad \nu_3(1) = \bar{1}, \quad \nu_3(2) = \bar{2},$$
$$\nu_3(3) = \bar{0}, \quad \nu_3(4) = \bar{1}, \quad \nu_3(5) = \bar{2},$$
$$\nu_3(6) = \bar{0}, \quad \nu_3(7) = \bar{1}, \quad \nu_3(8) = \bar{2}, \text{ etc.}$$

Since each $x \in Z$ is an element of exactly one coset $\bar{x}$, and since the set of cosets $\{\bar{0}, \bar{1}, \bar{2}, \ldots, \overline{n-1}\}$ is a partition of Z, ν_n is a surjection. Further, if $x, y \in G$, then

$$\nu_n(x + y) = \overline{x + y};$$

but (Theorem 27, Chapter 2, page 101,) we have proved that $\overline{x + y} = \bar{x} \oplus \bar{y}$. Hence

$$\nu_n(x + y) = \nu_n(x) \oplus \nu_n(y),$$

and so ν_n is an epimorphism.

Exercises

1. Let G and H be groups, and let $f : G \longrightarrow H$ be a homomorphism, and let $M < H$. Prove: If $f^{-1}[M] = \{x \mid x \in G \text{ and } f(x) \in M\}$, then $f^{-1}[M]$ is a subgroup of G.

2. If (Exercise 1) $M \lhd H$, prove that $f^{-1}[M] \lhd G$.

The type of epimorphism illustrated in Example 5 is of fundamental importance for group theory. We now define analogous kinds of homomorphisms for arbitrary groups. Let G be a group and let N be a normal subgroup of G. The elements of the quotient group G/N are cosets with respect to N and the set of all these cosets is a partition of G. Hence, if $x \in G$, then there is one and only one coset $\bar{x} = xN = Nx$ containing the element x. Now define $\nu_N : G \longrightarrow G/N$ by

$$(4) \qquad\qquad \nu_N(x) = \bar{x}$$

where $x \in G$ and $\bar{x}$ is the unique coset of N containing x. Further, if A is any coset of N, then A is a nonempty subset of G. If $y \in A$, then A is the only coset of N containing y and $\bar{y} = A$. Hence $\nu_N(y) = \bar{y} = A$. These remarks show that $\nu_N : G \longrightarrow G/N$ is a surjection.

THEOREM 6.   With G, N and ν_N as above, ν_N is an epimorphism.

Proof: We have already verified that ν_N is a surjection. Next let $x, y \in G$. Then

$$\nu_N(xy) = \overline{xy}$$
$$= \bar{x} * \bar{y} \quad \text{(Exercise 3, Chapter 2, page 115)}$$
$$= \nu_N(x) * \nu_N(y),$$

where $*$ is the group operation of G/N. Hence ν_N is an epimorphism.

Q.E.D.

DEFINITION 4.   The epimorphism $\nu_N\colon G \longrightarrow G/N$ defined by (4) is the *quotient* (or *natural* or *canonical*) epimorphism of G onto G/N.

Exercises

1. Prove that a composite of homomorphisms is a homomorphism. Given a set of groups, is epimorphism an equivalence relation on the set?

2. Let G and H be groups, and let $\eta\colon G \longrightarrow H$ be a homomorphism. Prove that $\eta[G]$ is a subgroup of H. Hence show that if e is the neutral element of G, then $\eta(e)$ is the neutral element of H.

3. Let $\eta\colon G \longrightarrow H$ be a group homomorphism, and let a be an element of finite order in G. Prove that the order of $\eta(a)$ divides the order of a.

4. Let G and H be groups and let $\eta\colon G \longrightarrow H$ and $\rho\colon H \longrightarrow G$ be epimorphisms such that $\rho \circ \eta$ is the identity mapping of G and $\eta \circ \rho$ is the identity mapping of H. Prove that η and ρ are isomorphisms such that $\rho = \eta^{-1}$.

6.   *A Relation Between Epimorphisms and Isomorphisms*

The importance of the quotient epimorphisms arises from the fact that all epimorphisms are obtainable from the quotient epimorphisms. The purpose of the present section is to establish this result; the meaning of "obtainable" will be clarified in the course of our investigation.

Suppose η is a homomorphism of a group G; if e is the neutral element of G, then $e' = \eta(e)$ is the neutral element of $G' = \eta[G]$. Hence the set

$$\{x \mid x \in G \text{ and } \eta(x) = e'\}$$

is nonempty.

DEFINITION 5.   The set $\{x \mid x \in G \text{ and } \eta(x) = e'\}$ is the *kernel* of the homomorphism η; it is denoted by "ker (η)."

THEOREM 7.   If η is a homomorphism of a group G, then ker (η) is a normal subgroup of G.

Proof: If e' is the neutral element of $\eta[G]$, then $\{e'\}$ is a normal subgroup of $\eta[G]$. Hence $\eta^{-1}[\{e'\}]$ is a normal subgroup of G (Exercise 2,

page 129). Since, clearly, ker $(\eta) = \eta^{-1}[\{e'\}]$, the theorem follows.
Q.E.D.

Theorem 7 establishes the fact that each homomorphism η of a group G determines a normal subgroup of G, namely, ker (η).

Exercises

1. Let η be a homomorphism of a group G with neutral element e. Prove that η is a monomorphism if and only if ker $(\eta) = \{e\}$.

2. With G and η as in Exercise 1 and $x, y \in G$, prove that $\eta(x) = \eta(y)$ if and only if x ker $(\eta) = y$ ker (η).

Suppose, now, a group G and a normal subgroup N of G are given. Does there exist a group H and epimorphism $\eta: G \longrightarrow H$ such that ker $(\eta) = N$? To answer this question, we recall Theorem 6. There we showed that given G and N, there exists an epimorphism, namely, $\nu_N: G \longrightarrow H = G/N$. If we can prove that

$$(5) \qquad\qquad \ker(\nu_N) = N,$$

we will be finished. But $x \in$ ker (ν_N) if and only if $\nu_N(x) = \bar{x} = xN$ is the neutral element of G/N, i.e., if and only if $xN = N$. However, since $xN = N$ if and only if $x \in N$, equation (5) is verified.

We come now to the final problem of this section. Given are multiplicative groups G and H and an epimorphism $\eta: G \longrightarrow H$. By Theorem 7, ker (η) is a normal subgroup of G, and by Theorem 6, the mapping $\nu_{ker(\eta)}: G \longrightarrow$ $G/$ker (η) defined by

$$\nu_{ker(\eta)}(x) = \bar{x}, \quad x \in G,$$

where $\bar{x} = x$ ker (η), is an epimorphism of G. Figure 9 illustrates the situation:

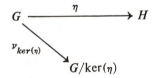

FIGURE 9

As we have seen, ker $(\nu_{ker(\eta)}) =$ ker (η). The question is:

What relation, if any, is there between $H = \eta[G]$ and the quotient group $G/\ker(\eta)$? The answer is given by

THEOREM 8.   There exists a unique isomorphism $\sigma: G/\ker(\eta) \to H$ such that

(6) $\sigma \circ \nu_{ker}(\eta) = \eta.$

In terms of a diagram, Theorem 8 means that the sketch in Figure 9 can be closed to form a triangle in exactly one way (Figure 10).

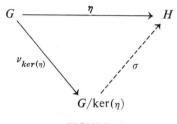

FIGURE 10

The important part of the theorem is the existence of the isomorphism σ; its uniqueness is a by-product of equation (6).

Proof: For each $x \in G$, set

(7) $\sigma(x \ker(\eta)) = \eta(x);$

we shall prove that σ is an isomorphism such that equation (6) holds. Equation (7) shows that the domain of σ is $G/\ker(\eta)$ and that its range is H. We show first that σ is an injection. Suppose $x, y \in G$ and $x \ker(\eta) = y \ker(\eta)$. Then $y^{-1}x \in \ker(\eta)$, hence (by definition of the kernel) $\eta(y^{-1}x) = e'$, the neutral element of H. Since η is an epimorphism,

$$\eta(x) = \eta(y);$$

hence

$$\sigma(x \ker(\eta)) = \sigma(y \ker(\eta)),$$

and therefore σ is a mapping. But, since each of the foregoing steps is reversible, it follows that σ is an injection.

Next, σ is an isomorphism. For,

$$
\begin{aligned}
\sigma(x \ker(\eta) \cdot y \ker(\eta)) &= \sigma((xy) \ker(\eta)) \\
&= \eta(xy) \qquad \text{(by equation (7))} \\
&= \eta(x)\eta(y) \ \text{(since η is a homorphism)} \\
&= \sigma(x \ker(\eta))\sigma(y \ker(\eta)).
\end{aligned}
$$

Finally, by definition of $\nu_{ker(\eta)}$, $\nu_{ker(\eta)}(x) = x \ker(\eta)$ for each

$x \in G$. Hence

$$\eta(x) = \sigma(x \ker(\eta))$$
$$= \sigma(\nu_{ker(\eta)}(x))$$
$$= (\sigma \circ \nu_{ker(\eta)})(x),$$

for each $x \in G$, and therefore

$$\sigma \circ \nu_{ker(\eta)} = \eta.$$

To establish the uniqueness of σ, suppose σ' is an isomorphism between $G/\ker(\eta)$ and H such that $\sigma' \circ \nu_{ker(\eta)} = \eta$. Then for each $\bar{x} \in G/\ker(\eta)$, $\bar{x} = x \ker(\eta) = \nu_{ker(\eta)}(x)$ for some $x \in G$, and

$$\sigma(\bar{x}) = \sigma(\nu_{ker(\eta)}(x)) = (\sigma \circ \nu_{ker(\eta)})(x) = \eta(x);$$

also

$$\sigma'(\bar{x}) = \sigma'(\nu_{ker(\eta)}(x)) = (\sigma' \circ \nu_{ker(\eta)})(x) = \eta(x).$$

Thus, for each $\bar{x} \in G/\ker(\eta)$,

$$\sigma(\bar{x}) = \sigma'(\bar{x});$$

therefore $\sigma = \sigma'$.

Q.E.D.

What is the significance of Theorem 8? Suppose G is a group and $H = \eta[G]$ is an epimorphic image of G. Then according to Theorem 8, we can obtain a "standard model"—more precisely, an isomorphic image—of H by constructing the quotient group $G/\ker (\eta)$. On the other hand, by Theorem 6, if N is a normal subgroup of G, then G/N is an epimorphic image (by the quotient epimorphism ν_N) of G. Thus all the "standard models" of the epimorphic images of G are obtainable by the process of constructing the qoutient groups G/N. In short:

If we know all the normal subgroups of G, we can construct "standard models" of all the epimorphic images of G.

Clearly, Theorem 8 may be regarded as a unique factorization theorem for epimorphisms. Given an epimorphism η of a group G, Theorem 8 shows that η is the product (i.e., composite) of the canonical epimorphism $G \longrightarrow G/\ker (\eta)$ and an isomorphism which is completely determined by η.

Finally, we remark that Theorem 8 has a minor generalization, as follows:

THEOREM 9. Let $\eta: G \longrightarrow H$ be a homomorphism. Then there exists a unique isomorphism σ such that

$$i \circ \sigma \circ \nu_{ker(\eta)} = \eta,$$

where i is the inclusion[2] mapping

$$i: (\sigma \circ \nu_{ker(\eta)}) [G] \longrightarrow H.$$

---

[2] If S is a subset of a set T, the mapping $i: S \longrightarrow T$ defined by $i(x) = x$, for each $x \in S$, is the *inclusion* mapping.

Diagrammatically, Theorem 9 is represented by Figure 11.

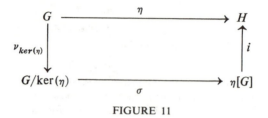

FIGURE 11

Proof: The proof of Theorem 9 is obtained from that of Theorem 8 by trivial modifications. It is left to the reader.

EXAMPLE 6.   We propose to exhibit an epimorphism $f: \mathscr{S}_4 \longrightarrow \mathscr{S}_3$, to determine ker($f$), and then to find an isomorphism $g: \mathscr{S}_4/\mathrm{ker}(f) \longrightarrow \mathscr{S}_3$ such that

$$g \circ \nu_{ker(f)} = f,$$

where $\nu_{ker(f)}$ is the quotient epimorphism $\nu_{ker(f)}: \mathscr{S}_4 \longrightarrow \mathscr{S}_4/\mathrm{ker}(f)$. These purposes are most easily served by making appropriate choices of the sets A_4 and A_3 on which $\mathscr{S}_4$ and $\mathscr{S}_3$, respectively, act.

We begin by choosing $A_4 = \{1, 2, 3, 4\}$; then the elements of $\mathscr{S}_4$ are (12), (13), ..., (123), (124), ..., (12)(34), (13)(24), (14)(23), etc. A_3 is now selected as *the subset of $\mathscr{S}_4$ consisting of the elements* (12)(34), (13)(24), (14)(23); thus

$$A_3 = \{(12)(34), (13)(24), (14)(23)\}.$$

Next we define $f: \mathscr{S}_4 \longrightarrow \mathscr{S}_3$ by means of the following device:

For each $\sigma \in \mathscr{S}_4$, set $f(\sigma) = \sigma'$ where σ' is defined by

(8) $\sigma'((i\,j)(k\;\ell)) = (\sigma(i)\;\sigma(j))(\sigma(k)\;\sigma(\ell)),$

and where i, j, k, ℓ are the four elements in A_4. To illustrate, if $\sigma = (1234)$, then

$$\sigma'((12)(34)) = (\sigma(1)\;\sigma(2))(\sigma(3)\;\sigma(4))$$
$$= (23)(41)$$
$$= (14)(23);$$

similarly,

$$\sigma'((13)(24)) = (\sigma(1)\;\sigma(3))(\sigma(2)\;\sigma(4))$$
$$= (24)(31)$$
$$= (13)(24),$$

and

$$\sigma'((14)(23)) = (\sigma(1)\;\sigma(4))(\sigma(2)\;\sigma(3))$$
$$= (21)(34)$$
$$= (12)(34).$$

Again, if $\sigma = (14)$, then

$$\sigma'((12)(34)) = (\sigma(1)\,\sigma(2))(\sigma(3)\,\sigma(4))$$
$$= (42)(31)$$
$$= (13)(24),$$
$$\sigma'((13)(24)) = (\sigma(1)\,\sigma(3))(\sigma(2)\,\sigma(4))$$
$$= (43)(21)$$
$$= (12)(34),$$

and

$$\sigma'((14)(23)) = (\sigma(1)\,\sigma(4))(\sigma(2)\,\sigma(3))$$
$$= (41)(23)$$
$$= (14)(23).$$

To improve his dexterity with equation (8) and the definition of f, the reader should determine σ' for several other values of σ, say, $\sigma = (12)(34)$, $(13)(24)$, $(14)(23)$.

We claim: If $\sigma \in \mathscr{S}_4$, then the corresponding $\sigma' = f(\sigma)$ is an element in $\mathscr{S}_3$. Indeed, if $(ij)(k\,\ell) \in A_3$, then since σ is one-one, $\sigma'((ij)(k\,\ell)) = (\sigma(i)\,\sigma(j))(\sigma(k)\,\sigma(\ell)) \in A_3$. The reader can verify easily that $\sigma': A_3 \longrightarrow A_3$ is a bijection, hence that $\sigma' \in \mathscr{S}_3$.

Next, f is a mapping. For, if $\sigma, \tau \in \mathscr{S}_4$ and $\sigma = \tau$, then for each $i \in A_4$, $\sigma(i) = \tau(i)$, and so

$$\sigma'((ij)(k\,\ell)) = (\sigma(i)\,\sigma(j))(\sigma(k)\,\sigma(\ell))$$
$$= (\tau(i)\,\tau(j))(\tau(k)\,\tau(\ell))$$
$$= \tau'((ij)(k\,\ell)).$$

Consequently $\sigma' = \tau'$ and f is a mapping.

Further, f is a homomorphism. If $\sigma, \tau \in \mathscr{S}_4$, then

$$(\sigma\tau)'((ij)(k\,\ell)) = (\sigma\tau(i)\,\sigma\tau(j))(\sigma\tau(k)\,\sigma\tau(\ell))$$
$$= (\sigma(\tau(i))\,\sigma(\tau(j)))(\sigma(\tau(k))\,\sigma(\tau(\ell)))$$
$$= \sigma'((\tau(i)\,\tau(j))(\tau(k)\,\tau(\ell)))$$
$$= \sigma'(\tau'((ij)(k\,\ell)))$$
$$= (\sigma'\tau')((ij)(k\,\ell)),$$

for each $(ij)(k\,\ell) \in A_3$. Hence $(\sigma\tau)' = \sigma'\tau'$ and f is a homomorphism.

Finally, f is a surjection. For convenience, set

$$x_1 = (14)(23),\ x_2 = (13)(24),\ x_3 = (12)(34).$$

If $\sigma = (12) \in \mathscr{S}_4$, then

$$\sigma'(x_1) = \sigma'((14)(23)) = (24)(13) = x_2,$$
$$\sigma'(x_2) = x_1,$$
$$\sigma'(x_3) = x_3,$$

and therefore $\sigma' = (x_1 \, x_2)$. In exactly the same way, if

$$\sigma = \text{identity (in } \mathcal{S}_4), \text{ then } \sigma' = \text{identity (in } \mathcal{S}_3);$$
$$\sigma = (13), \text{ then } \sigma' = (x_1 x_3);$$
$$\sigma = (23), \text{ then } \sigma' = (x_2 x_3);$$
$$\sigma = (123), \text{ then } \sigma' = (x_1 x_2 x_3);$$
$$\sigma = (132), \text{ then } \sigma' = (x_1 x_3 x_2).$$

Thus each element in $\mathcal{S}_3$ is an image, under f, of some element in $\mathcal{S}_4$ and so f is surjective.

What is ker(f)? A permutation $\sigma \in \mathcal{S}_4$ is in ker(f) if and only if σ' is the identity mapping of A_3. In particular, $\sigma'(x_3) = x_3$, i.e.,

$$\sigma'((12)(34)) = (\sigma(1)\,\sigma(2))(\sigma(3)\,\sigma(4)) = (12)(34).$$

Hence either

(9) $(\sigma(1)\,\sigma(2)) = (12) \text{ and } (\sigma(3)\,\sigma(4)) = (34)$

or else

(10) $(\sigma(1)\,\sigma(2)) = (34) \text{ and } (\sigma(3)\,\sigma(4)) = (12).$

If equations (9) hold, the alternatives for σ are:

identity, $\sigma = (12)(34)$, $\sigma = (34)$, or $\sigma = (12)$.

The last two alternatives are ruled out since

$$(34)'(x_1) \neq x_1 \text{ and } (12)'(x_1) \neq x_1.$$

On the other hand, it is clear that (1) and (12)(34) are in ker (f). In a similar fashion, we deduce from equations (10) that the remaining elements in ker(f) are (13)(24) and (14)(23). Therefore

$$\text{ker}(f) = \{(1), (12)(34), (13)(24), (14)(23)\} = \mathcal{F}_4,$$

the *Four Group*. By Theorem 7, $\mathcal{F}_4 = \text{ker}(f)$ is a normal subgroup of $\mathcal{S}_4$. If, as usual, $\nu_{\mathcal{F}_4} = \nu_{\text{ker}(f)}$ is the quotient epimorphism $\mathcal{S}_4 \longrightarrow \mathcal{S}_4/\mathcal{F}_4$, then by Theorem 8 there is a unique isomorphism $g: \mathcal{S}_4/\mathcal{F}_4 \longrightarrow \mathcal{S}_3$ such that $g \circ \nu_{\mathcal{F}_4} = f$.

Exercises

1. In Example 6, determine $g(\sigma \mathcal{F}_4)$ for each $\sigma \in \mathcal{S}_4$.

2. Let Z be the additive group of integers and let $G = (x)$ be a finite cyclic group of order m. Define $f: Z \longrightarrow G$ by $f(n) = x^n, n \in Z$. (i) Prove that f is an epimorphism; (ii) show that $\text{ker}(f) = mZ = \{my \,|\, y \in Z\}$. Hence, using Theorem 8, deduce that G and the modular arithmetic Z_m are isomorphic.

3. Let A and N be subgroups of a group G such that

for each $a \in A$ and each $n \in N$, $a^{-1}na \in N$.

Let $AN = \{an \,|\, a \in A \text{ and } n \in N\}$. Prove: (i) AN is a subgroup of G; (ii) N is a normal subgroup of AN, and (iii) $A \cap N$ is a normal subgroup of A.

4. With A, N and G as in Exercise 3, define $f(an) = a(A \cap N)$ for each $a \in A$ and each $n \in N$. Prove:
(a) $f: AN \longrightarrow A/(A \cap N)$ is a surjection;
(b) f is an epimorphism;
(c) $\ker(f) = N$.
(d) Hence deduce from Theorem 8 that $AN/N \cong A/(A \cap N)$. This result is the *First Isomorphism Theorem*.

5. Let N and H be normal subgroups of G such that $N \subset H \subset G$, and define $g(xN) = xH$ for each $x \in G$. Prove:
(a) $g: G/N \longrightarrow G/H$ is an epimorphism;
(b) $\ker(g) = H/N$, hence $H/N \lhd G/N$.
(c) Hence deduce from Theorem 8 that $(G/N)/(H/N) \cong G/H$.
This is the *Second Isomorphism Theorem*. The First and Second Isomorphism Theorems are fundamental for the further study of group theory.

Exercises 6-12 below, together with Exercise 3, provide proofs of the two isomorphism theorems without resorting to Theorem 8. Beginning with Exercise 6 we assume that G_1 and G_2 are groups with normal subgroups H_1 and H_2, respectively. Further, we let $\iota_1: H_1 \longrightarrow G_1$ and $\iota_2: H_2 \longrightarrow G_2$ be the injection homomorphisms (i.e., $\iota_n(x) = x$, $x \in H_n$, $n = 1, 2$), and $\nu_1: G_1 \longrightarrow G_1/H_1$ and $\nu_2: G_2 \longrightarrow G_2/H_2$ be the quotient epimorphisms. Our hypotheses are summarized in Figure 12.

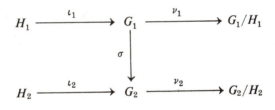

FIGURE 12

Prove the following:
6. There exists a unique mapping $\bar{\sigma}: G_1/H_1 \longrightarrow G_2/H_2$ such that

(11) $$\bar{\sigma} \circ \nu_1 = \nu_2 \circ \sigma,$$

if and only if $\sigma[H_1] \subset H_2$. If such a $\bar{\sigma}$ exists, it is a homomorphism.
7. There exists a unique mapping $\sigma': H_1 \longrightarrow H_2$ such that

(12) $$\iota_2 \circ \sigma' = \sigma \circ \iota_1,$$

if and only if $\sigma[H_1] \subset H_2$. If such a σ' exists, it is a homomorphism.
Recapitulating Exercises 6 and 7, we have

THEOREM 10. A necessary and sufficient condition for the existence of unique mappings $\bar{\sigma}$ or σ' satisfying (11) and (12), respectively, is that $\sigma[H_1] \subset H_2$. In particular, if one of the two mappings exists so does the other, and both are homomorphisms.

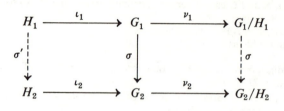

FIGURE 13

Now assume that the condition $\sigma[H_1] \subset H_2$ is satisfied. Prove:

8. A necessary and sufficient condition that $\bar{\sigma}$ be a surjection is that $\nu_2 \circ \sigma$ be a surjection.

9. A necessary and sufficient condition that $\bar{\sigma}$ be one-one is that $\sigma^{-1}[H_2] = H_1$.

10. Discover and verify analogous, necessary and sufficient conditions that (i) σ' be one-one, (ii) σ' be a surjection.

11. (*First Isomorphism Theorem*) Let A and N be subgroups of G satisfying the conditions of Exercise 3. Thus

$$A \cap N \longrightarrow A \longrightarrow A/A \cap N$$
$$\sigma \downarrow$$
$$N \longrightarrow AN \longrightarrow AN/N,$$

where σ is the injection homomorphism. Use the results of Exercises 6–10 to prove that $A/A \cap N \cong AN/N$.

12. (*Second Isomorphism Theorem*) With N, H and G as in Exercise 5,

$$H \longrightarrow G \longrightarrow G/H$$
$$\nu_N \downarrow$$
$$H/N \longrightarrow G/N \longrightarrow (G/N)/(H/N),$$

use Exercises 6–10 to prove

$$(G/N)/(H/N) \cong G/H.$$

7. *Endomorphisms of a Group*

This final section on group homomorphisms is devoted to a brief study of homomorphisms $\sigma: G \longrightarrow H$ where $H = G$. Such homomorphisms have numerous uses not only in algebra, but in many mathematical disciplines which use algebraic techniques and ideas (for an example, see E. Artin, *Geometric Algebra*, Chapter II, New York, Interscience Publishers, Inc., 1957).

DEFINITION 6.   A homomorphism $\eta: G \longrightarrow G$ is an *endomorphism* of G; an isomorphism $\eta: G \longrightarrow G$ is an *automorphism* of G.

The set of all endomorphisms of G is denoted by "$\mathscr{E}(G)$"; the set of all automorphisms of G is denoted by "$\mathscr{A}(G)$."

EXAMPLE 7.   In Example 6 it is proved that there is an epimorphism of $\mathscr{S}_4$ to $\mathscr{S}_3$ where $\mathscr{S}_4$ is the set of all permutations of $\{1, 2, 3, 4\}$ and $\mathscr{S}_3$ is the set of all permutations of $\{1, 2, 3\}$. Let

$$\mathscr{S}_3' = \{\varphi \mid \varphi \in \mathscr{S}_4 \text{ and } \varphi(4) = 4\};$$

then

$$\mathscr{S}_3' = \{(1), (12), (13), (23), (123), (132)\},$$

and clearly $\mathscr{S}_3 \cong \mathscr{S}_3'$. Hence there exists an epimorphism $\mathscr{S}_4 \longrightarrow \mathscr{S}_3' < \mathscr{S}_4$ which is an endomorphism of $\mathscr{S}_4$.

EXAMPLE 8.   If G is a group with neutral element e, then $\omega : G \longrightarrow G$ defined by

$$\omega(x) = e, \quad x \in G,$$

is clearly an endomorphism. And $\iota : G \longrightarrow G$ defined by

$$\iota(x) = x, \quad x \in G,$$

is evidently an automorphism. The mappings ω and ι are the *trivial* endomorphisms.

EXAMPLE 9.   The purpose of this example is to exhibit (a) an endomorphism $\eta : G \longrightarrow G$ which is surjective, yet is not an automorphism, and (b) a monomorphism $\mu : G \longrightarrow G$ such that $\mu[G] \neq G$, hence is not an automorphism.

Recall that a *sequence of integers* is a mapping

$$\sigma : N \longrightarrow Z$$

where N is the set of all nonnegative integers; thus for each $n \in N$, $\sigma(n) = a_n \in Z$. For convenience, we write

$$\sigma = (a_0, a_1, a_2, \ldots).$$

If σ, τ are sequences of integers, define $\sigma \oplus \tau$ by

$$(\sigma \oplus \tau)(n) = \sigma(n) + \tau(n), \quad n \in N.$$

In other words, if $\sigma = (a_0, a_1, a_2, \ldots)$ and $\tau = (b_0, b_1, b_2, \ldots)$, then

$$\sigma \oplus \tau = (a_0 + b_0, a_1 + b_1, a_2 + b_2, \ldots).$$

Let G be the set of all sequences of integers; with $\oplus$ as operation, it is easy to see that G is a group. Now define $\eta : G \longrightarrow G$ by

$$\eta((a_0, a_1, a_2, \ldots)) = (a_1, a_2, \ldots).$$

Clearly η is an endomorphism. It is not a monomorphism since, for example,

$$\eta((0, a_1, a_2, \ldots)) = \eta((1, a_1, a_2, \ldots)).$$

On the other hand, η is a surjection. Indeed, if $\tau = (b_0, b_1, b_2, \ldots) \in G$, then $\tau' = (0, b_0, b_1, b_2, \ldots) \in G$ and $\eta(\tau') = \tau$.

The mapping $\mu : G \longrightarrow G$ defined by

$$\mu((a_0, a_1, a_2, \ldots)) = (0, a_0, a_1, a_2, \ldots)$$

is a monomorphism, but since there is no $\sigma \in G$ such that $\mu(\sigma) = (1, 0, 0, \ldots)$, μ is not an automorphism.

In Exercise 7, page 123, it is proved that for each element a in the group G, the mapping $\sigma_a : G \longrightarrow G$ defined by

$$\sigma_a(x) = axa^{-1}, \quad x \in G,$$

is an automorphism of G.

DEFINITION 7.   For each $a \in G$, the automorphism σ_a is the *inner automorphism determined by a*. An automorphism which is not inner is *outer*. The set of inner automorphisms of G is denoted by "$\mathscr{I}(G)$."

THEOREM 11.   For each group G, $\mathscr{A}(G)$ is a group with respect to composition of mappings as group operation. Moreover, $\mathscr{I}(G)$ is a normal subgroup of $\mathscr{A}(G)$.

Proof: The first statement is left to the reader to prove.

To show that $\mathscr{I}(G)$ is a group, note first that $\mathscr{I}(G) \neq \emptyset$. Next, let $\sigma_a, \sigma_b \in \mathscr{I}(G)$, where $a, b \in G$. Then for each $x \in G$,

$$(\sigma_a \circ \sigma_b)(x) = \sigma_a(\sigma_b(x)) = \sigma_a(bxb^{-1}) = a(bxb^{-1})a^{-1}$$
$$= (ab)x(ab)^{-1} = \sigma_{ab}(x),$$

whence $\sigma_a \circ \sigma_b \in \mathscr{I}(G)$. Further,

$$(\sigma_{a^{-1}} \circ \sigma_a)(x) = \sigma_{a^{-1}}(axa^{-1}) = a^{-1}(axa^{-1})(a^{-1})^{-1} = x, \quad x \in G;$$

thus $\sigma_{a^{-1}} = \sigma_a^{-1}$ and, by definition, $\sigma_{a^{-1}} \in \mathscr{I}(G)$. By Theorem 20, Chapter 2, page 138, $\mathscr{I}(G) < \mathscr{A}(G)$.

Finally, to prove that $\mathscr{I}(G) \lhd \mathscr{A}(G)$, it suffices to verify that if $\tau \in \mathscr{A}(G)$ and $\sigma_a \in \mathscr{I}(G)$, then $\tau \circ \sigma_a \circ \tau^{-1} \in \mathscr{I}(G)$. For each $x \in G$,

$$(\tau \circ \sigma_a \circ \tau^{-1})(x) = (\tau \circ \sigma_a)(\tau^{-1}(x))$$
$$= \tau(\sigma_a(\tau^{-1}(x)))$$
$$= \tau(a\tau^{-1}(x)a^{-1})$$
$$= \tau(a)\tau(\tau^{-1}(x))\tau(a^{-1})$$
$$= \tau(a)x(\tau(a))^{-1}$$
$$= \sigma_{\tau(a)}(x);$$

therefore $\tau \circ \sigma_a \circ \tau^{-1} = \sigma_{\tau(a)}$ and so $\tau \circ \sigma_a \circ \tau^{-1} \in \mathscr{I}(G)$.

Q.E.D.

We conclude our brief discussion of endomorphisms by describing some important properties of the set $\mathscr{E}(G)$ in the case that G is an (additive) Abelian group. For all $\xi, \eta \in \mathscr{E}(G)$ we define a sum $\oplus$ by

(13) $(\xi \oplus \eta)(x) = \xi(x) + \eta(x), \quad x \in G.$

Since $\xi(x), \eta(x)$ are elements in G, it follows that $\xi \oplus \eta$ is a mapping $G \longrightarrow G$.

But further, $\xi \oplus \eta$ is a homomorphism. For, if $x, y \in G$, then

$$(\xi \oplus \eta)(x + y) = \xi(x + y) + \eta(x + y)$$
$$= (\xi(x) + \xi(y)) + (\eta(x) + \eta(y))$$
$$= (\xi(x) + \eta(x)) + (\xi(y) + \eta(y))$$
$$= (\xi \oplus \eta)(x) + (\xi \oplus \eta)(y);$$

thus $\xi \oplus \eta$ is a homomorphism and so $\xi \oplus \eta \in \mathscr{E}(G)$. The reader should now verify that $\oplus$ is commutative and associative.

Next let $\omega: G \to G$ be the mapping defined by

$$\omega(x) = 0, \quad x \in G;$$

if $x, y \in G$, then $x + y \in G$ and therefore

$$\omega(x + y) = 0 = 0 + 0 = \omega(x) + \omega(y).$$

Thus ω is a homomorphism, whence $\omega \in \mathscr{E}(G)$. Now if $\xi \in \mathscr{E}(G)$, define $-\xi$ by

$$(-\xi)(x) = -(\xi(x)), \quad x \in G.$$

Evidently $-\xi$ is a mapping $G \to G$. And if $x, y \in G$, then

$$(-\xi)(x + y) = -(\xi(x + y))$$
$$= -(\xi(x) + \xi(y))$$
$$= -\xi(x) + (-(\xi(y)))$$
$$= (-\xi)(x) + (-\xi)(y),$$

so that $-\xi \in \mathscr{E}(G)$. Moreover, for each $x \in G$,

$$((-\xi) \oplus \xi)(x) = (-\xi)(x) + \xi(x)$$
$$= -(\xi(x)) + \xi(x)$$
$$= 0$$
$$= \omega(x);$$

thus for each $\xi \in \mathscr{E}(G)$ there is a $-\xi \in \mathscr{E}(G)$ such that $(-\xi) \oplus \xi = \omega$. To sum up, we have proved

THEOREM 12. *If G is an additive Abelian group, then $\mathscr{E}(G)$ is a group with respect to $\oplus$ (defined by equation (13)). The neutral element of $\mathscr{E}(G)$ is the trivial endomorphism ω.*

Exercises

1. Let G be an additive, Abelian group. (a) Prove: If $\xi, \eta \in \mathscr{E}(G)$, then the composite $\xi \circ \eta$ is an element in $\mathscr{E}(G)$. (b) If $\xi, \eta, \zeta \in \mathscr{E}(G)$, prove that $\xi \circ (\eta \oplus \zeta) = \xi \circ \eta \oplus \xi \circ \zeta$ and $(\eta \oplus \zeta) \circ \xi = \eta \circ \xi \oplus \zeta \circ \xi$.

2. Let η be an endomorphism of the additive group of the integers Z. Prove that there is an integer k such that

(14) $\eta(x) = kx, \quad x \in Z.$

Conversely, show that for each $k \in Z$, (14) defines an endomorphism of the additive group of Z.

3. Prove that $\mathscr{E}(Z)$ is isomorphic with the additive group of Z. (*Hint*: Each element in $\mathscr{E}(Z)$ is an η_k where

$$\eta_k(x) = kx, \quad x \in Z,$$

and k is an integer. Define $\varphi: Z \longrightarrow \mathscr{E}(Z)$ by $\varphi(k) = \eta_k$ and show that φ is an isomorphism.)

4. Let G be a finite cyclic group $|G| = n \geq 1$. Prove that the group $\mathscr{E}(G)$ is isomorphic with the modular arithmetic Z_n.

5. Let G be a group with neutral element e. Prove that G is Abelian if and only if $\mathscr{I}(G) = \{e\}$.

4

The Theory of Rings

1. Introduction

Rings are mathematical structures with two binary operations. As such, one may expect them to be more complex than groups. However, we shall see that the group theory of Chapters 2 and 3 provides some information concerning rings and also some tools and guides for their investigation.

The theory of rings arises in a natural way out of a study of the familiar number systems of elementary arithmetic and algebra. Indeed, the first examples of rings are the well-known structures:

> I. the system of all integers,
> II. the system of all rational numbers,
> III. the system of all real numbers,
> IV. the system of all complex numbers.

One may regard each of I-III as included in its successor. However, comparisons other than inclusion can be found among I-IV. All four systems have certain properties in common, and each has certain properties which distinguish it from the others. What are some of the properties that the systems I-IV share? First of all, addition and multiplication are defined within each system; and the results of these operations applied to numbers within a given system yield numbers which are again in that system. For instance, the sum of two integers (rationals, real numbers, complex numbers) is again an integer (rational, real number, complex number). An analogous statement is true for multiplication. These facts may appear more noteworthy when we realize that there are systems of which such statements are not true. For example, the system consisting of all the odd integers does *not* have the property that the sum of each two of its numbers is again a member of the system. Thus $3 + 5 = 8$ is an even number. A second illustration is the system consisting of all the negative integers, $\{-1, -2, -3, \cdots\}$. A product of two negative numbers is positive, hence is *not* a member of the system.

A further property that the systems I-IV have in common is commutativity of addition; the rule

$$a + b = b + a$$

is valid for the integers, the rationals, the real numbers, and the complex numbers. We now list in a formal way several properties that the systems I-IV have in common.

1. Each system possesses a binary operation of addition and a binary operation of multiplication; i.e., for each pair a, b in the system there are uniquely defined elements $a + b$ and ab in the system.
2. Addition is *commutative*, $a + b = b + a$.
3. Addition is *associative*, $a + (b + c) = (a + b) + c$.
4. There is an element 0 such that $0 + a = a$.
5. For each element a there is an element b such that $b + a = 0$.
6. Multiplication is *commutative*, $ab = ba$.
7. Multiplication is *associative*, $a(bc) = (ab)c$.
8. Multiplication is *distributive* over addition, $a(b + c) = ab + ac$.

Let R be any one of the four systems I-IV. Then properties 2-5 comprise the assertion: R is an additive, Abelian group. Hence 1-8 may be summarized by:

R is a system with two binary operations $+$ and $\cdot$ —addition and multiplication—such that

1'. R is an additive, Abelian group.
2'. Multiplication is commutative.
3'. Multiplication is associative.
4'. Multiplication is distributive over addition, i.e., $a(b + c) = ab + ac$ for all $a, b, c \in R$.

Are I-IV the only systems with two binary operations having properties 1'-4'? No! As one might expect, there are (infinitely) many others. For instance, consider

EXAMPLE 1.   *The modular arithmetics, Z_n.* To fix our ideas we focus upon Z_6. In accordance with Section 14, Chapter 2, addition in Z_6 is given by the table:

TABLE 7

$\oplus$	$\bar{0}$	$\bar{1}$	$\bar{2}$	$\bar{3}$	$\bar{4}$	$\bar{5}$
$\bar{0}$	$\bar{0}$	$\bar{1}$	$\bar{2}$	$\bar{3}$	$\bar{4}$	$\bar{5}$
$\bar{1}$	$\bar{1}$	$\bar{2}$	$\bar{3}$	$\bar{4}$	$\bar{5}$	$\bar{0}$
$\bar{2}$	$\bar{2}$	$\bar{3}$	$\bar{4}$	$\bar{5}$	$\bar{0}$	$\bar{1}$
$\bar{3}$	$\bar{3}$	$\bar{4}$	$\bar{5}$	$\bar{0}$	$\bar{1}$	$\bar{2}$
$\bar{4}$	$\bar{4}$	$\bar{5}$	$\bar{0}$	$\bar{1}$	$\bar{2}$	$\bar{3}$
$\bar{5}$	$\bar{5}$	$\bar{0}$	$\bar{1}$	$\bar{2}$	$\bar{3}$	$\bar{4}$

In an (apparently) artificial way we now define a multiplication, $\odot$, by means of the table:

TABLE 8

$\odot$	$\bar{0}$	$\bar{1}$	$\bar{2}$	$\bar{3}$	$\bar{4}$	$\bar{5}$
$\bar{0}$	$\bar{0}$	$\bar{0}$	$\bar{0}$	$\bar{0}$	$\bar{0}$	$\bar{0}$
$\bar{1}$	$\bar{0}$	$\bar{1}$	$\bar{2}$	$\bar{3}$	$\bar{4}$	$\bar{5}$
$\bar{2}$	$\bar{0}$	$\bar{2}$	$\bar{4}$	$\bar{0}$	$\bar{2}$	$\bar{4}$
$\bar{3}$	$\bar{0}$	$\bar{3}$	$\bar{0}$	$\bar{3}$	$\bar{0}$	$\bar{3}$
$\bar{4}$	$\bar{0}$	$\bar{4}$	$\bar{2}$	$\bar{0}$	$\bar{4}$	$\bar{2}$
$\bar{5}$	$\bar{0}$	$\bar{5}$	$\bar{4}$	$\bar{3}$	$\bar{2}$	$\bar{1}$

The reader should try to discover the rationale governing the construction of Table 8, and apply it to other modular arithmetics.

It is easy to check that Z_6, with multiplication defined above, has properties 1'–4'. However, the multiplication in this arithmetic exhibits certain anomalies when compared with the standard arithmetics I-IV. For instance, in Table 8 we see that

$$\bar{2} \odot \bar{3} = \bar{0} \quad \text{and} \quad \bar{3} \odot \bar{4} = \bar{0};$$

these results are striking when one realizes that $\bar{0}$ corresponds to the number zero in the familiar number systems.

We now turn our attention to another simple, modular arithmetic Z_2. Addition in Z_2 is given by

TABLE 9

$\oplus$	$\bar{0}$	$\bar{1}$
$\bar{0}$	$\bar{0}$	$\bar{1}$
$\bar{1}$	$\bar{1}$	$\bar{0}$

and multiplication is defined by

TABLE 10

$\odot$	$\bar{0}$	$\bar{1}$
$\bar{0}$	$\bar{0}$	$\bar{0}$
$\bar{1}$	$\bar{0}$	$\bar{1}$

In this case it is trivial to check the validity of 1'–4'.

Now, by equation (22), Chapter 2, page 102,

$$\bar{0} = 0 + 2Z = 2Z = \{2n \mid n \in Z\},$$

and

$$\bar{1} = 1 + 2Z = \{1 + 2n \mid n \in Z\}.$$

Thus $\bar{0}$ is merely the set of all even integers and $\bar{1}$ is the set of all odd integers.

If we replace "$\bar{0}$" and "$\bar{1}$" by "even" and "odd," respectively, throughout Tables 9 and 10 we obtain

TABLE 11

$\oplus$	even	odd
even	even	odd
odd	odd	even

and

TABLE 12

$\odot$	even	odd
even	even	even
odd	even	odd

Tables 11 and 12 summarize the knowledge essential to those computations which depend only upon the evenness and oddness of integers. For this reason, Z_2 with multiplication defined by Table 10 (or, Table 12) is called the "arithmetic of even and odd."

Evidently, there is a vast array of systems satisfying 1'-4'. If we are to study such systems in detail we have two alternative procedures:

(i) Either we continue to seek, in a more or less haphazard way, systems which satisfy 1'-4' and to investigate each system as we find it, or else

(ii) we deduce a number of consequences of 1'-4' and find out what utility this additional knowledge possesses.

If we follow the latter path, the consequences deduced will be valid in *every* system satisfying 1'-4'. Thus a great deal can be learned about such systems before we have even found them. And moreover, our deductions may (in fact, will) lead to general methods for the construction of large classes of systems satisfying 1'-4'. Abstract algebra pursues the second alternative and so do we.

Before embarking upon our deductive journey we remark that the abstract approach does *not* free us of the obligation to examine specific systems satisfying 1'-4'. On the contrary, concrete examples are needed to develop intuition and to suggest particular theorems and even general lines of investigation.

2. *Definition of Ring*

DEFINITION 1. A *ring* is a set R on which are defined two binary operations $+$ and $\cdot$ (*addition* and *multiplication*, respectively) such that

(1) R is an Abelian group with respect to $+$. The neutral element of this group is denoted by "0," *zero*;

(2) the multiplication, $\cdot$, is associative;

(3) the multiplication is distributive over addition, i.e., for all a, b, c in R, $a(b + c) = ab + ac$ and $(b + c)a = ba + ca$.

Systems I-IV, as well as the two arithmetics Z_6 and Z_2, of Example 1, are instances of rings.

We emphasize that although in our initial examples of rings, multiplication is commutative, Definition 1 does *not* require commutativity of the product. Indeed, there are important rings in which multiplication is noncommutative and the theory of noncommutative rings is very extensive.

EXAMPLE 2. *A noncommutative ring.*[1] Let $M_2 = Z \times Z \times Z \times Z$; thus M_2 consists of all ordered quadruples (a_1, a_2, a_3, a_4), $a_i \in Z$, $1 \leq i \leq 4$. To facilitate the definition of the binary operations on M_2, the ordered quadruples (a_1, a_2, a_3, a_4) are usually written

$$\begin{pmatrix} a_1 & a_2 \\ a_3 & a_4 \end{pmatrix}$$

and are called "2×2 matrices over the integers." Define $+$ and $\cdot$ on M_2 by

$$\begin{pmatrix} a_1 & a_2 \\ a_3 & a_4 \end{pmatrix} + \begin{pmatrix} b_1 & b_2 \\ b_3 & b_4 \end{pmatrix} = \begin{pmatrix} a_1 + b_1 & a_2 + b_2 \\ a_3 + b_3 & a_4 + b_4 \end{pmatrix}$$

and

$$\begin{pmatrix} a_1 & a_2 \\ a_3 & a_4 \end{pmatrix}\begin{pmatrix} b_1 & b_2 \\ b_3 & b_4 \end{pmatrix} = \begin{pmatrix} a_1b_1 + a_2b_3 & a_1b_2 + a_2b_4 \\ a_3b_1 + a_4b_3 & a_3b_2 + a_4b_4 \end{pmatrix},$$

respectively. It is immediate from the definitions that $+$ and $\cdot$ are binary operations on M_2 and that with respect to $+$, M_2 is an Abelian group whose neutral element is $\begin{pmatrix} 0 & 0 \\ 0 & 0 \end{pmatrix}$. With a little more work it can be shown that $\cdot$ is associative and distributive over $+$. However, if a_2 or a_3 is not zero, then

$$\begin{pmatrix} a_1 & a_2 \\ a_3 & a_4 \end{pmatrix}\begin{pmatrix} 1 & 0 \\ 0 & 0 \end{pmatrix} = \begin{pmatrix} a_1 & 0 \\ a_3 & 0 \end{pmatrix}$$

and

$$\begin{pmatrix} 1 & 0 \\ 0 & 0 \end{pmatrix}\begin{pmatrix} a_1 & a_2 \\ a_3 & a_4 \end{pmatrix} = \begin{pmatrix} a_1 & a_2 \\ 0 & 0 \end{pmatrix}$$

show that multiplication is not commutative. Thus M_2 is a noncommutative ring with respect to $+$ and $\cdot$; it is the *ring of 2 $\times$ 2 matrices over the integers.*

---

[1] To avoid disrupting the flow, it might be well to skip Example 2 in a first reading of this chapter.

Exercises

1. Does the set of nonnegative integers with ordinary addition and multiplication form a ring? Why?

2. Let $G = \{a + bi \mid a, b \in Z$ and $i^2 = -1\}$. Define addition and multiplication by

$$(a + bi) + (c + di) = (a + c) + (b + d)i$$

and

$$(a + bi) \cdot (c + di) = (ac - bd) + (ad + bc)i,$$

respectively. Show that G is a ring (the ring of *Gaussian integers*) with respect to $+$ and $\cdot$

3. Which of the following structures form rings with respect to ordinary addition and multiplication: $\{a + b\sqrt{2} \mid a, b \in Z\}$; the set of all even integers; the set of all integral multiples of -3; for each integer k, the set of all integral multiples of k?

4. Let A be an Abelian group and let $\mathscr{E}(A)$ be the set of endomorphisms of A. Prove that if addition and multiplication are defined on $\mathscr{E}(A)$ as in equation (13), chapter 3, page 140, and in Exercise 1, page 141, respectively, then $\mathscr{E}(A)$ is a ring. Exhibit a group A for which the multiplication in $\mathscr{E}(A)$ is noncommutative.

3. *Some Properties of Rings*

In this section we deduce a few of the simplest properties of rings. Among the results to be established is the famous theorem,

$$(1) \qquad\qquad\qquad (-1)(-1) = 1.$$

In high school algebra courses, one often resorts to some remarkable contortions to convince students of the validity of equation (1). It will be seen that the steps required to deduce this result from the definition of ring are simple enough to be understandable to high school students.

Theorems 1–5 below are merely restatements of results obtained in Chapter 2, this time restricted to the case of additive, Abelian groups. The proofs are therefore omitted. Theorems 6 and 7 are new and involve the second binary operation, multiplication. Throughout, R is a ring with binary operations $+$ and $\cdot$ (or, juxtaposition).

THEOREM 1. A ring R has exactly one zero element.

THEOREM 2. For each $a \in R$, there is one and only one $b \in R$ such that $b + a = a + b = 0$.

DEFINITION 2. If $a \in R$, the element $b \in R$ such that $b + a = 0$ is the *additive inverse* of a and is denoted "$-a$."

THEOREM 3. For each $a \in R$, $-(-a) = a$.

THEOREM 4. For all a, x, y in R, $x = y$ if and only if $a + x = a + y$. Also $x = y$ if and only if $x + a = y + a$.

THEOREM 5. For each a and each b in R, there is a unique $x \in R$ such that $b = a + x = x + a$.

THEOREM 6. For each $a \in R$, $a \cdot 0 = 0 \cdot a = 0$.

Proof: Since 0 is the neutral element of the additive group of R, $0 + a = a$ and so

$$(0 + a) \cdot a = a \cdot a.$$

But $(0 + a) \cdot a = 0 \cdot a + a \cdot a$ and therefore

$$0 \cdot a + a \cdot a = a \cdot a = 0 + a \cdot a.$$

Therefore by Theorem 4, $0 \cdot a = 0$.

In a similar manner, it is proved that $a \cdot 0 = 0$. This additional proof is needed since we have not assumed that multiplication is commutative. The details are left to the reader.

Q.E.D.

THEOREM 7. For each a and each b in R,

$$(-a)(-b) = ab.$$

Proof: We know that each element in R has exactly one additive inverse in R. Therefore if we can show that $(-a)(-b)$ and ab are additive inverses of the same element in R, it must follow that $(-a)(-b) = ab$.

Now

$$(-a)(-b) + a(-b) = ((-a) + a)(-b) \quad \text{(distributive law)}$$
$$= 0(-b)$$
$$= 0, \quad \text{(Theorem 6)}$$

hence $(-a)(-b)$ is the additive inverse of $a(-b)$. But

$$ab + a(-b) = a(b + (-b))$$
$$= a0$$
$$= 0,$$

so ab is also the additive inverse of $a(-b)$ and Theorem 7 is proved.

Q.E.D.

Our definition of ring demands the presence of a neutral element 0 for addition, but makes no such requirement with respect to multiplication. Indeed, there are rings (e.g., the ring of even integers) having no neutral element for multiplication. Nevertheless, rings possessing a neutral element for multiplication are of special importance—witness the usual arithmetic systems.

> **DEFINITION 3.** An element e in a ring R such that $ea = ae = a$ for each $a \in R$ is a *unity element* for R. If R possesses a unity element, it is said to be a "ring with unity element."

The definition of unity element requires *both* $ae = a$ and $ea = a$, for each $a \in R$. It is impossible to derive one of these conditions from the other. This can be demonstrated by exhibiting a ring, necessarily noncommutative, having, say, a *left* unity element which is not a *right* unity element.

> **EXAMPLE 3.** Let R be the set of all 2×2 matrices $\begin{pmatrix} a & b \\ 0 & 0 \end{pmatrix}$, where a and b are integers. With respect to the matrix addition and multiplication defined in Example 2, page 147, R is a ring. The matrix $\begin{pmatrix} 1 & 0 \\ 0 & 0 \end{pmatrix} \in R$ is a left unity element for R since
>
> $$\begin{pmatrix} 1 & 0 \\ 0 & 0 \end{pmatrix}\begin{pmatrix} a & b \\ 0 & 0 \end{pmatrix} = \begin{pmatrix} a & b \\ 0 & 0 \end{pmatrix};$$
>
> however, it is not a right unity element since
>
> $$\begin{pmatrix} a & b \\ 0 & 0 \end{pmatrix}\begin{pmatrix} 1 & 0 \\ 0 & 0 \end{pmatrix} = \begin{pmatrix} a & 0 \\ 0 & 0 \end{pmatrix}.$$

Exercises

Prove:

1. If R is a ring with unity element, then R has only one unity element. (*Hint:* Follow the proof of Theorem 1, Chapter 2.)

2. In a ring R with unity element e, $(-e)(-e) = e$. In the familiar arithmetic systems, this boils down to $(-1)(-1) = 1$.

3. For each a and each b in a ring R, $(-a)b = a(-b) = -(ab)$.

4. $-0 = 0$.

5. In a ring containing exactly two (exactly three) elements, multiplication is commutative.

6. There exists a ring containing exactly four elements in which multiplication is noncommutative. This result is proved by exhibiting such a ring.

7. Give an example of a ring which has a right unity element that is not a left unity element.

4. *The Modular Arithmetics, Again*

The two rings discussed in Example 1 (page 144) were modular arithmetics, each augmented by an appropriate multiplication. In this section we shall describe a uniform method for introducing a multiplication in every modular arithmetic. The multiplications so obtained will (i) convert each Z_n into a ring, and (ii) in the case of Z_6 and Z_2, specialize to the multiplications defined by Tables 8 and 10, respectively. In the following we shall use freely the concepts and results in Sections 13 and 14 of Chapter 2.

To begin, we require,

THEOREM 8. If a, b, c, d are integers such that $a \equiv c$ (mod n) and $b \equiv d$ (mod n), $n \in N$, then

$$ab \equiv cd \ (\text{mod } n).$$

Proof: The proof is very easy. By Theorem 24, Chapter 2, page 98, $a \equiv c$ (mod n) implies $ab \equiv bc$ (mod n). And similarly, $b \equiv d$ (mod n) implies $bc \equiv cd$ (mod n). Since $\equiv$ (mod n) is transitive, the desired result follows.

Q.E.D.

If, as in Section 14, Chapter 2, we let $\bar{x}$ be the residue class of x (mod n), i.e.,

$$\bar{x} = \{y \,|\, y \in Z \text{ and } y \equiv x \ (\text{mod } n)\},$$

then Theorem 8 yields the

COROLLARY. If $\bar{a}, \bar{b}, \bar{c}, \bar{d}$ are residue classes in Z_n such that $\bar{a} = \bar{c}$ and $\bar{b} = \bar{d}$, then

$$\overline{ab} = \overline{cd}.$$

Proof: By hypothesis, we know that

$$\left. \begin{array}{c} a \equiv c \\ b \equiv d \end{array} \right\} (\text{mod } n).$$

Hence $ab \equiv cd$ (mod n) and so

$$\overline{ab} = \overline{cd}.$$

Q.E.D.

We now make the important

DEFINITION 4. Let $\bar{a}$ and $\bar{b}$ be residue classes in Z_n. The *product* $\bar{a} \odot \bar{b}$ of $\bar{a}$ and $\bar{b}$ is defined as the residue class $\overline{ab}$; thus

(2)
$$\bar{a} \odot \bar{b} = \overline{ab}.$$

The corollary to Theorem 8 shows that $\odot$ is a mapping, $\odot: Z_n \times Z_n \longrightarrow Z_n$, and therefore it is a binary operation on Z_n. Note that since $\bar{x} = x + nZ$, where $x \in Z$, the defining equation (2) can be rewritten as

$$(a + nZ) \odot (b + nZ) = ab + nZ.$$

Exercises

1. Check that Tables 8 and 10 are deducible from Definition 4.
2. Compute multiplication tables for Z_5, Z_7, Z_8 and Z_9.

THEOREM 9. If multiplication is defined by equation (2), then the modular arithmetic Z_n is a commutative ring with unity element.

Proof: In Theorem 27, Chapter 2, page 101, we showed that Z_n is an Abelian group with respect to $\oplus$. Next, we verify that conditions (2) and (3) of Definition 1 are satisfied.

Multiplication is associative. For, if $\bar{a}, \bar{b}, \bar{c} \in Z_n$, then

$$\begin{aligned}
(\bar{a} \odot \bar{b}) \odot \bar{c} &= \overline{ab} \odot \bar{c} \\
&= \overline{(ab)c} \\
&= \overline{a(bc)} \\
&= \bar{a} \odot \overline{bc} \\
&= \bar{a} \odot (\bar{b} \odot \bar{c}).
\end{aligned}$$

Further, multiplication is distributive over addition. Indeed,

$$\begin{aligned}
\bar{a} \odot (\bar{b} \oplus \bar{c}) &= \bar{a} \cdot \overline{b + c} \\
&= \overline{a(b + c)} \\
&= \overline{ab + ac} \\
&= \overline{ab} \oplus \overline{ac} \\
&= \bar{a} \odot \bar{b} \oplus \bar{a} \odot \bar{c}.
\end{aligned}$$

The proofs that $\odot$ is commutative and that Z_n has a unity element are left to the reader.

Q.E.D.

In case $n > 1$ the unity element of Z_n is $\bar{1} = 1 + nZ$; in case $n = 1$ the unity element is $\bar{0}$.

Hitherto, in speaking of the modular arithmetic Z_n, we have had in mind the Abelian group Z_n with group operation $\oplus$. From now on we shall adopt the convention set forth in

DEFINITION 5. The *modular arithmetic with modulus n* is the ring Z_n with binary operations $\oplus$ and $\odot$.

The earlier concept of modular arithmetic henceforth will be called "the additive group of Z_n."

The reader now has examined, in some detail, several specific modular arithmetics, namely, Z_2, Z_5, Z_6, Z_7, Z_8 and Z_9. In each of Z_6, Z_8 and Z_9 there are residue classes $\bar{a}$ and $\bar{b}$ such that $\bar{a} \neq \bar{0}$, $\bar{b} \neq \bar{0}$, but $\bar{a} \odot \bar{b} = \bar{0}$. For instance, in Z_6, $\bar{2} \odot \bar{3} = \bar{0}$; in Z_8, $\bar{2} \odot \bar{4} = \bar{0}$; in Z_9, $\bar{3} \odot \bar{3} = \bar{0}$. By contrast, in Z_2, Z_5 and Z_7 there are no such residue classes.

DEFINITION 6. Let R be a ring. An element $a \in R$, $a \neq 0$, is a *left divisor of zero* if and only if there is an element $b \in R$, $b \neq 0$, such that $ab = 0$. Similarly, $c \in R$, $c \neq 0$, is a *right divisor of zero* if and only if there is an element $d \in R$, $d \neq 0$, such that $dc = 0$. A nonzero element in R is a *divisor of zero* if and only if it is both a left and right divisor of zero.

Obviously, in a commutative ring every left (right) divisor of zero is a divisor of zero. Thus, in Z_6, $\bar{2}$ and $\bar{3}$ are divisors of zero; in Z_8, $\bar{2}$ and $\bar{4}$ are divisors of zero; Z_9 has only one divisor of zero, namely, $\bar{3}$. On the other hand, in noncommutative rings there may be left (right) divisors of zero which are not right (left) divisors of zero.

EXAMPLE 4. Let R be the set of all 2×2 matrices $\begin{pmatrix} a & 0 \\ b & 0 \end{pmatrix}$ where $a, b \in Z$, and let the ring operations in R be the ordinary matrix addition and multiplication. In this ring $\begin{pmatrix} 1 & 0 \\ 0 & 0 \end{pmatrix}$ is a left divisor of zero, since

$$\begin{pmatrix} 1 & 0 \\ 0 & 0 \end{pmatrix}\begin{pmatrix} 0 & 0 \\ 1 & 0 \end{pmatrix} = \begin{pmatrix} 0 & 0 \\ 0 & 0 \end{pmatrix}.$$

However, $\begin{pmatrix} 1 & 0 \\ 0 & 0 \end{pmatrix}$ is not a right divisor of zero since

$$\begin{pmatrix} a & 0 \\ b & 0 \end{pmatrix}\begin{pmatrix} 1 & 0 \\ 0 & 0 \end{pmatrix} = \begin{pmatrix} a & 0 \\ b & 0 \end{pmatrix},$$

and $\begin{pmatrix} a & 0 \\ b & 0 \end{pmatrix} \neq \begin{pmatrix} 0 & 0 \\ 0 & 0 \end{pmatrix}$ unless both $a = 0$ and $b = 0$, i.e., unless $\begin{pmatrix} a & 0 \\ b & 0 \end{pmatrix}$ is the zero matrix.

The reader should give an example of a ring containing a right divisor of zero which is not a left divisor of zero.

We ask: Under what conditions will a modular arithmetic have divisors of zero and when will it fail to have any? In the modular arithmetics examined thus far, those for which the modulus is a prime number contain no divisors of zero, while those for which the modulus is composite do contain divisors of zero. This leads us to conjecture

THEOREM 10. A modular arithmetic contains divisors of zero if and only if its modulus n is composite.

Proof: Suppose n is composite, say, $n = q_1q_2$ where q_1 and q_2 are integers such that $1 < q_1 < n$ and $1 < q_2 < n$. Then $\bar{q}_1 = q_1 + nZ$, $\bar{q}_2 = q_2 + nZ$ and $\bar{q}_1 \neq \bar{0}$, $\bar{q}_2 \neq \bar{0}$. Therefore

$$\bar{q}_1 \odot \bar{q}_2 = q_1q_2 + nZ = n + nZ = nZ = \bar{0},$$

and so $\bar{q}_1$ and $\bar{q}_2$ are divisors of zero.

Conversely, suppose n is prime and let $\bar{r}, \bar{s}$ be residue classes such that $\bar{r} \odot \bar{s} = \bar{0}$. If $\bar{r} = r + nZ$ and $\bar{s} = s + nZ$, then

$$\bar{r} \odot \bar{s} = rs + nZ.$$

Hence by our hypothesis, $rs + nZ = \bar{0} = nZ$, and so $rs \in nZ$. Therefore $n|rs$; but since n is prime, $n|r$ or $n|s$. If, say, $n|r$, then $r = qn$ and $\bar{r} = qn + nZ = nZ = \bar{0}$. Similarly, if $n|s$, then $\bar{s} = \bar{0}$. Thus in a modular arithmetic Z_n with prime modulus n, $\bar{r} \odot \bar{s} = \bar{0}$ implies $\bar{r} = \bar{0}$ or $\bar{s} = \bar{0}$. Hence Z_n contains no divisors of zero.

Q.E.D.

5. Integral Domains

Henceforth, unless the contrary is stated explicitly, every ring under consideration is assumed to be commutative.[2]

We have observed that some rings have divisors of zero and others do not.

DEFINITION 7. A ring with unity element which contains no divisors of zero is an *integral domain*.

Thus the integers, the rationals, the real numbers, the complex numbers, and all modular arithmetics with prime moduli are integral domains; no modular arithmetic with composite modulus is an integral domain.

One of the important properties of multiplication that holds in integral domains but not in rings with divisors of zero is the *cancellation law for multiplication*.

---

[2] In some of the theorems below the hypothesis of commutativity is not needed. This assumption is useful because it enables us to avoid making a special hypothesis in a few cases. *It will be most instructive for the reader to check each theorem from now on and determine whether or not the proof holds up without the assumption of commutativity.*

THEOREM 11.   (The Cancellation Law for Multiplication). Let a, b, c be elements in an integral domain R. If $ab = ac$ and $a \neq 0$, then $b = c$.

> *Proof:* If $ab = ac$, then $ab - ac = 0$ and therefore $a(b - c) = 0$. But since $a \neq 0$ and since an integral domain contains no divisors of zero, it follows that $b - c = 0$, whence $b = c$.
>
> Q.E.D.

Conversely,

THEOREM 12.   If R is a ring in which the cancellation law for multiplication holds, then R contains no divisors of zero. Hence if R has a unity element, then R is an integral domain.

> *Proof:* Suppose $ab = 0$ where $a \neq 0$. Using the cancellation law for multiplication, we shall prove that $b = 0$, hence a is not a divisor of zero. Indeed,
>
> $$ab = 0 = a0;$$
>
> and $ab = a0$, $a \neq 0$, implies by the cancellation law that $b = 0$.
>
> Q.E.D.

Putting Theorems 11 and 12 together, it is clear that an integral domain may be defined either as:
a) a ring with unity element containing no divisors of zero, *or* as
b) a ring with unity element in which the cancellation law for multiplication holds.

6.   *Fields*

An important difference among the various rings is this: In the ring of rationals, for each rational number $a \neq 0$ there is a *reciprocal* or *multiplicative inverse* $1/a$ such that $1/a \cdot a = 1$. The same is true for the reals, the complex numbers and, as we shall see, the modular arithmetics with prime moduli. On the other hand, in certain other rings, there exist nonzero elements not having a reciprocal in the given ring. Thus in the ring (in fact, integral domain) of integers, the integer 2 does not possess a reciprocal. In the modular arithmetic with modulus 6, $\bar{3}$ does not possess a multiplicative inverse.

Exercises

1. Why cannot the zero element in a nontrivial ring possibly have a reciprocal?
2. Find all the elements in Z_3, Z_5, Z_6, Z_7, Z_8 and Z_9, respectively, having

multiplicative inverses. Does any pattern emerge? If not, do the same for Z_{10}, Z_{11}, Z_{12}, and again answer the question.

Those rings in which each nonzero element possesses a reciprocal in the given ring constitute an important class of rings.

DEFINITION 8. A *field* F is a commutative ring with unity element $e \neq 0$ such that for each nonzero $a \in F$ there is a *multiplicative inverse* or *reciprocal* $b \in F$ with the property that

$$ba = e.$$

That $ab = e$ follows from the commutativity of multiplication. If b is the reciprocal of a, one writes "a^{-1}" in place of b.

Exercises

1. Prove: For each $a \neq 0$ in a field F, there is one and only one multiplicative inverse.

2. Prove that $F - \{0\}$ is a multiplicative group.

THEOREM 13. A field contains no divisors of zero. Hence every field is an integral domain.

Proof: Suppose a and b are elements in a field F where $ab = 0$ and $a \neq 0$; we show that $b = 0$, hence a is not a divisor of zero. Since $a \neq 0$ and F is a field, there is an $a^{-1} \in F$ such that $a^{-1}a = e$. Then

$$b = eb = (a^{-1}a)b = a^{-1}(ab) = a^{-1}0 = 0.$$

Q.E.D.

Rings, integral domains and fields may be thought of as forming a tower of structures of which the class of rings forms the base, the class of integral domains the next layer, and the class of fields the topmost layer. This picture reflects the fact that every integral domain is a ring and every field is an integral domain. We shall explore other relationships among these classes as the text develops.

Exercise 2, page 155, suggests the following

THEOREM 14. If p is a prime number, then the modular arithmetic Z_p is a field. Conversely, if Z_n is a field, then n is prime.

Proof: First recall that $\bar{1} = 1 + pZ$ is the unity element of Z_p. Now let $\bar{a} \in Z_p$, $\bar{a} \neq \bar{0}$; we show that $\bar{a}$ has a reciprocal in Z_p and therefore Z_p is a field. Define $\sigma_{\bar{a}} : Z_p \longrightarrow Z_p$ by

$$\sigma_{\bar{a}}(\bar{x}) = \bar{x} \odot \bar{a};$$

clearly $\bar{x} = \bar{y}$ if and only if $\bar{x} \odot \bar{a} = \bar{y} \odot \bar{a}$ (details?). And since

Z_p is a finite set, it follows that σ_a is a bijection. Consequently there is a $\bar{v} \in Z_p$ such that

$$\sigma_a(\bar{v}) = \bar{v} \odot \bar{a} = \bar{1}$$

and so $\bar{v}$ is a reciprocal of $\bar{a}$.

The converse is left to the reader.

Q.E.D.

Exercises

1. Let e be an element in an integral domain D, and let $a \in D$, $a \neq 0$. Prove: If $ae = a$, then e is a unity element for D.

2. Prove: Every finite integral domain containing at least two elements is a field. (*Hint:* Use the technique of proof of Theorem 14 and Exercise 1.)

To reassure the reader that the generalization of Theorem 14 embodied in Exercise 2 above is significant, we give an example of a finite integral domain containing more than one element which is *not* a modular arithmetic.

EXAMPLE 5. Let $D = \{0, e, a, b\}$; addition and multiplication are defined for D by the following tables:

TABLE 13

+	0	e	a	b
0	0	e	a	b
e	e	0	b	a
a	a	b	0	e
b	b	a	e	0

TABLE 14

·	0	e	a	b
0	0	0	0	0
e	0	e	a	b
a	0	a	b	e
b	0	b	e	a

The table for · shows that D contains no divisors of zero. And with a finite amount of labor, one can verify that D is a ring. Hence D is a finite integral domain with e as unity element.

How can we be sure that D is not one of the modular arithmetics? This is easy. Indeed, for each n, the modular arithmetic Z_n contains exactly n elements, namely, $\bar{0}, \bar{1}, \cdots, \overline{n-1}$. Since D contains exactly four elements, if D were one of the Z_n, it would have to be Z_4. But Z_4 contains a divisor of zero, namely, $\bar{2}$, whereas D has none. Hence $D \neq Z_4$ and therefore D is not a modular arithmetic.

THEOREM 15. If F is a field, then for all a, b in F, where $a \neq 0$, there is a unique $x \in F$ such that

$$ax = b.$$

Proof: Existence. Take $x = a^{-1}b$; then

$$a(a^{-1}b) = (aa^{-1})b = eb = b.$$

Uniqueness. Suppose $ay = b$; then $a^{-1}(ay) = a^{-1}b$, whence $(a^{-1}a)y = a^{-1}b$ or $ey = y = a^{-1}b$.

Q.E.D.

Exercises

1. In the field Z_3, find $\bar{1}/\bar{2}$ (i.e., the reciprocal of $\bar{2}$).
2. In Z_5, find $\bar{1}/\bar{2}$, $\bar{1}/\bar{3}$, $\bar{1}/\bar{4}$, $\bar{2}/\bar{3}$, $\bar{3}/\bar{4}$, $\bar{4}/\bar{7}$, $\bar{8}/\bar{3}$.
3. Is there a $\bar{1}/\bar{2}$ in Z_2? a $\bar{1}/\bar{3}$ in Z_3? a $\bar{1}/\bar{5}$ in Z_5?
4. For each integer $n > 1$, does there exist a field with exactly n elements? Why? (*Hint:* See if there is a field with exactly six elements. But before starting off, list all commutative groups of order six.)

Suppose R is a ring with unity element e; by definition, R is an Abelian group with respect to $+$, and therefore (Definition 5′, page 64) mx is defined for each integer $m > 0$ and for all $x \in R$ as

$$mx = \overbrace{x + x + \cdots + x}^{m \text{ times}}.$$

Now a curious property exhibited by the modular arithmetics—as well as other rings—but not shared by the number systems of elementary mathematics, is this: There exists an integer m such that

$$m\bar{1} = \overbrace{\bar{1} \oplus \bar{1} \oplus \cdots \oplus \bar{1}}^{m \text{ times}} = \bar{0}.$$

For instance, in Z_5, $5\,\bar{1} = \bar{1} \oplus \bar{1} \oplus \bar{1} \oplus \bar{1} \oplus \bar{1} = \bar{0}$; in Z_2, $\bar{1} \oplus \bar{1} \oplus \bar{1} \oplus \bar{1} = \bar{0}$, etc. Similarly, in the finite field D, of Example 5, we see that $2e = e + e = 0$. On the other hand, for the ring of integers (as well as the rationals, the reals and the complex numbers), there is no integer $m > 0$ such that $m1 = \overbrace{1 + 1 + \cdots + 1}^{m \text{ times}} = 0$. To facilitate the classification of rings relative to this property, we introduce

DEFINITION 9. Let R be a ring with unity element e. If there is an integer $n > 0$ such that

$$ne = \overbrace{e + e + \cdots + e}^{n \text{ times}} = 0,$$

then R is a ring with *finite characteristic*. If there is no such integer n, then R is a ring with *characteristic zero*.

Suppose R is a ring with finite characteristic. Then the set

$$K = \{n \mid n \in Z, n > 0, \text{ and } ne = 0\}$$

is nonempty. By the well-ordering principle, K contains a smallest positive integer m.

DEFINITION 10.   If R is a ring with finite characteristic, then the smallest positive integer m such that

$$me = 0$$

is the *characteristic* of R, and R is a *ring of characteristic m*. One writes

$$\text{char } R = m.$$

Theorems 16 and 17 below are proved in an informal, hand-waving fashion. Formal proofs, by induction, are left to the reader.

THEOREM 16.   If char $R = m$, then $mx = 0$ for each $x \in R$.

Proof: By definition, $mx = \overbrace{x + x + \cdots + x}^{m \text{ times}}$, for each $x \in R$. On the other hand, if e is the unity element of R, then

$$ex = x,$$

for each $x \in R$, whence

$$mx = \overbrace{ex + ex + \cdots + ex}^{m \text{ times}}$$
$$= \overbrace{(e + e + \cdots + e)}^{m \text{ times}}x \qquad \text{(Distributive law)}$$
$$= (me)x \qquad\qquad \text{(Definition of } me).$$

Since char $R = m$, $me = 0$; since $0x = 0$, we have

$$mx = 0x = 0.$$

Q.E.D.

THEOREM 17.   For each $n \in Z$, $n > 0$, char $Z_n = n$.

Proof: Since $\bar{1}$ is the unity element of Z_n, we show that (a) $n\bar{1} = \bar{0}$; and (b) if $0 < m < n$, then $m\bar{1} \neq \bar{0}$.
First,

$$n\bar{1} = \overbrace{\bar{1} \oplus \bar{1} \oplus \cdots \oplus \bar{1}}^{n \text{ times}}$$
$$= \overbrace{(1 + nZ) \oplus (1 + nZ) \oplus \cdots \oplus (1 + nZ)}^{n \text{ times}}$$
$$= \overbrace{(1 + 1 + \cdots + 1)}^{n \text{ times}} + nZ$$
$$= n + nZ$$
$$= nZ$$
$$= \bar{0},$$

hence (a) holds. Next, let m be a positive integer such that $m\bar{1} = \bar{0}$. Then, as above,

$$m\bar{1} = m + nZ = nZ;$$

but $m + nZ = nZ$ implies $m \in nZ$. Hence $n \mid m$ and so either $m = 0$ or $m \geq n$. Therefore (b) follows.

Q.E.D.

The only fields among the Z_n are those for which the modulus n is prime, and by Theorem 17, each of these fields has prime characteristic. In fact, Theorem 17 shows that the modulus and the characteristic are equal. However, there are fields other than the Z_p, p prime, having finite characteristic, e.g., the ring of Example 5. What can we say about the characteristic of such fields?

THEOREM 18. If F is a field with characteristic $m \neq 0$, then m is a prime number.

Proof: Suppose m is composite, say, $m = qr$ where $1 < q < m$ and $1 < r < m$. Since m is the smallest positive integer such that $me = 0$ (e is the unity element of F), $re \neq 0$ and $qe \neq 0$. But

$$\begin{aligned}
(re)(qe) &= (rq)e \qquad \text{(why?)} \\
&= me \\
&= 0,
\end{aligned}$$

and therefore re and qe are divisors of zero in F. This conclusion contradicts Theorem 13. Hence m is prime.

Q.E.D.

Exercises

For the purposes of Exercises 1–5, define

$$\binom{n}{k} = \frac{n!}{(n-k)!\,k!},$$

where n and k are integers and $0 \leq k \leq n$.

1. Prove: If $k > 1$, then $\binom{n}{k-1} + \binom{n}{k} = \binom{n+1}{k}$.

2. Prove: In a ring R,

$$(a + b)^n = \sum_{k=0}^{n} \binom{n}{k} a^{n-k} b^k, \qquad a, b \in R.$$

(Use the PFI.)

3. Prove: If n and k are integers, $0 \leq k \leq n$, then $\binom{n}{k}$ is an integer.

4. Prove: If n is prime, then $\binom{n}{k}$, $0 < k < n$, is divisible by n.

5. Prove: If R is a ring with unity element having prime characteristic $n > 0$, then $(a + b)^n = a^n + b^n$.

6. Let R be a ring containing exactly m elements, $m > 0$. Prove that $ma = 0$ for each $a \in R$.

7. Is Theorem 18 valid for integral domains? If so, prove it. If not, give a counterexample.

7. Subrings

Having classified a variety of rings, we turn next to a study of ring structure. A basic concept is that of subring.

DEFINITION 11. Let R be a commutative ring with binary operations $+$ and $\cdot$ and let S be a subset of R. Further, let $\oplus$ and $\odot$ be the restrictions of $+$ and $\cdot$, respectively, to S. If S is a ring with respect to $\oplus$ and $\odot$, then S is a *subring* of R, and R is an *extension ring* of S.

If S is a subring of R, then with respect to addition, S is a subgroup of the additive group of R (why?) and therefore $S \neq \emptyset$. It is also useful to define:

(a) If S is a subring of R, and if S is an integral domain (a field), then S is a *subdomain* (a *subfield*) of R; and correspondingly,

(b) If S is a subring of R and R is an integral domain (a field), then R is an *extension domain* (an *extension field*) of S.

EXAMPLE 6. Let R be a ring and let $S = \{0\}$; then $\emptyset \neq S \subset R$ and since $0 \oplus 0 = 0 + 0 = 0$ and $0 \odot 0 = 0 \cdot 0 = 0$, the restrictions $\oplus$ and $\odot$ of $+$ and $\cdot$, respectively, to S are binary operations on S. For the rest, it is trivial to check that S is a ring with respect to $\oplus$ and $\odot$.

One may also take $S = R$ and in this case $\oplus = +$ and $\odot = \cdot$, and clearly S is a ring with respect to $\oplus$ and $\odot$. Thus R is a subring of itself. The subrings $\{0\}$ and R, of R, are usually regarded as the *trivial* subrings; all others are the *proper* subrings.

EXAMPLE 7. The integers may be considered as a subdomain of the rationals, the rationals are a subfield of the reals, and the reals are a subfield of the field of complex numbers.

EXAMPLE 8. Let R be the modular arithmetic Z_{12}, and let $S = \{\bar{0}, \bar{4}, \bar{8}\}$. It is easy to check that S is a subring of Z_{12}.

Exercises

1. Find all the proper subrings of Z_4, Z_6, Z_8, Z_9.

2. Do the rings Z_2, Z_3, Z_5, Z_7 have proper subrings?

3. Prove: If R is a ring such that $|R| = m$ and if S is a subring of R such that $|S| = k$, then $k \mid m$.

4. Prove: The fields Z_p, where p is a prime number, contain no proper subrings.

5. Prove: If R is an integral domain and if S is a subring of R containing the unity element of R, then S is an integral domain.

6. Suppose R is a field and S is a subring of R. Does it follow that S is a subfield of R? Why?

To determine from the definition whether or not a given subset S of a ring R is a subring of R, one would have to go to the trouble of verifying that the restrictions of $+$ and $\cdot$ to S are binary operations on S, and that S together with these restrictions satisfy the conditions of Definition 1. Much of this labor can be avoided if one uses

THEOREM 19. A nonempty subset S of a ring R is a subring of R if and only if

    (i) for each a and each b in S, $a - b \in S$, and

    (ii) for each a and each b in S, $ab \in S$.

Note: The binary operations in (i) and (ii) are *not* the restrictions of $+$ and $\cdot$ to S; they are the *original* binary operations on R.

    Proof: Suppose, first, S is a subring of R. Then with respect to addition, S is a subgroup of R (hence $S \neq \emptyset$), and so for each a and each b in S, $a - b \in S$ (Theorem 21, Chapter 2, 90). Thus condition (i) holds. Next, let $\odot$ be the restriction of $\cdot$ S. Then $\odot$ is a binary operation on S and therefore if $a, b \in S$, then $a \odot b \in S$. But $ab = a \odot b$; hence $ab \in S$ and so (ii) holds.

    Conversely, let S be a nonempty subset of R satisfying (i) and (ii) and let $\oplus$ and $\odot$ be the restrictions of $+$ and $\cdot$, respectively. We prove that S is a ring with respect to $\oplus$ and $\odot$, hence S is a subring of R.

    By Theorem 21, Chapter 2, S is a subgroup of the additive group of R, therefore S is a group with respect to $\oplus$. And by Theorem 18, Chapter 2, 86, $\odot$ is a binary operation on S. To complete the proof that S is a ring with binary operations $\oplus$ and $\odot$, we verify:

    (a) $\odot$ is associative. For, if $a, b, c \in S$, then

$$a \odot (b \odot c) = a(bc) = (ab)c = (a \odot b) \odot c;$$

    (b) $\odot$ is distributive over $\oplus$. Again, with $a, b, c \in S$ we have

$$a \odot (b \oplus c) = a(b + c) = ab + ac = a \odot b \oplus a \odot c.$$

A similar argument shows that $(b \oplus c) \odot a = b \odot a \oplus c \odot a$, and the proof of Theorem 19 is complete.

Q.E.D.

If R is a commutative ring, then it is easy to see that S is likewise commutative. In this case only one distributive law need be verified in S.

The reader should reexamine Examples 6, 7 and 8 in the light of Theorem 19.

Suppose S is a subring of a ring R. It is customary to drop the notations "$\oplus$" and "$\odot$" for the restrictions of $+$ and $\cdot$, respectively, to S. Instead, one uses the symbols "$+$" and "$\cdot$" to denote both the original binary operations on R as well as their restrictions to S. Wherever no confusion is likely to result, we shall abide by this convention.

Exercises

1. Prove: If n is an integer, then $nZ = \{nk \mid k \in Z\}$ is a subring of Z.

2. (Converse of Exercise 1) Prove: If S is a subring of Z, then there exists an integer $n \geq 0$ such that $S = nZ$. (*Hint:* Examine the discussion of Example 19, Chapter 2, page 91.)

3. (i) Let S and T be subrings of a ring R. Prove that $S \cap T$ is a subring of R. (ii) If $\mathscr{S}$ is a set of subrings of R, prove that $\cap \mathscr{S}$ is a subring of R.

Remark: Suppose A is a subset of a ring R, and $\mathscr{S} = \{S \mid S$ is a subring of R and $A \subset S\}$. By Exercise 3, part (ii), $T = \cap \mathscr{S}$ is a subring of R and clearly $A \subset T$. T is the *subring of R generated by A*. The reader will note that the subring of R generated by A is the *smallest* subring of R containing A as a subset (details?).

4. (i) Let $A = \{1, \sqrt{2}\}$ be a subset of the field R of real numbers, and let $[A]$ be the set of all finite sums and differences of products

$$x_1^{n_1} x_2^{n_2} \cdots x_k^{n_k},$$

where (a) $k \geq 1$, (b) each $x_i \in A$, (c) each n_i is a positive integer. Prove that $[A]$ is the subring of R generated by A. Also show that a real number α is in $[A]$ if and only if there are integers a and b such that $\alpha = a + b\sqrt{2}$.

(ii) Let R be a ring with unity element, let $x \in R$, and let $A = \{x\}$. Define the ring $[A]$ by analogy with (i) above and prove that $[A]$ is the subring of R generated by the set $\{x\}$. Give a simple description of the elements of $[A]$.

5. Let A be a subset of a ring R with unity element. If $A = \emptyset$, set $[A] = \{0\}$. If $A \neq \emptyset$, define $[A]$ by analogy with Exercise 4, part (i). Prove that $[A]$ is the subring of R generated by A.

6. Let T be a subset of a ring R, and let

$$\mathscr{S} = \{S \mid S \text{ is a subring of } R \text{ and } T \subset S\}.$$

Prove that $\cap \mathscr{S} = [T]$.

7. Let A be a subring of a ring R. Prove that $[A] = A$.

8. Let Q be the field of rational numbers. Describe the elements of the subring $[Q \cup \{\pi\}]$, of $R(=$ reals$)$.

9. Let A be a subring of the ring R, and let $t \in R$. Describe the elements of the subring $[A \cup \{t\}]$ generated by the set $A \cup \{t\}$.

10. Let $R = Z \times Z$ and define $+$ and $\cdot$ on R by

(3) $$(a, b) + (c, d) = (a + c, b + d)$$

and

(4) $$(a, b) \cdot (c, d) = (ac, bd),$$

respectively. *Note:* The right hand $+$ and $\cdot$ are not the same as the left hand $+$ and $\cdot$ in equations (3) and (4). (i) Prove that R is a ring with respect to $+$ and $\cdot$ having $(1, 1)$ as unity element. (ii) Let $S = \{(a, b) \mid (a, b) \in R$ and $b = 0\}$; prove that S is a subring of R and its unity element is $(1, 0)$. Hence the unity element of a subring need *not* be the unity element of the extension ring.

11. Let $R = Z \times 2Z$ and define $+$ and $\cdot$ on R by equations (3) and (4), respectively. (i) Prove that R is a ring. Does R have a unity element? (ii) Let $S = \{(a, b) \mid (a, b) \in R$ and $b = 0\}$. Prove that $(1, 0)$ is the unity element of S. Thus a subring may have a unity element but the extension ring need not have one.

12. Let S be a subring of R, and suppose e' is the unity element of S and e is the unity element of R. Prove: If $e' \neq e$, then e' is a divisor of zero in R.

8.   *Ring Homomorphisms*

From Section 7 we see that the subring concept in ring theory corresponds to the subgroup concept in group theory. To press the analogy further, we should seek the kind of subring corresponding to the *normal* subgroup. One way of approaching this problem is to determine the subrings which play a role in ring theory, similar to that of normal subgroups in group theory. In Chapter 3, Theorem 7, 130 (and the remarks on page 131), we saw that the normal subgroups of a group are precisely the kernels of the group homomorphisms. This observation suggests that we begin by defining *ring homomorphisms* and then proceed to study the kernels of these homomorphisms.

DEFINITION 12.   Let R be a ring with respect to operations $+$ and $\cdot$, and let T be a set on which are defined binary operations $\oplus$ and $\odot$. Further, let $\eta : R \longrightarrow T$ be a mapping.

(i) η is a *ring homomorphism* if and only if for all $x, y \in R$,

(5) $$\eta(x + y) = \eta(x) \oplus \eta(y)$$

and

(6) $$\eta(xy) = \eta(x) \odot \eta(y).$$

Equations (5) and (6) comprise the *morphism property* of η.

Assuming η is a ring homomorphism, we define further:

(ii) η is a *ring epimorphism* if and only if it is a surjection;

(iii) η is a *ring monomorphism* if and only if it is one-one;

(iv) η is a *ring isomorphism* if and only if it is both an epimorphism and a monomorphism.

If η is a ring homomorphism (alternatively, monomorphism) then $\eta[R] = \{\eta(x) \mid x \in R\}$ is the *epimorphic image* (alternatively, *isomorphic image*) of R under η.

Wherever it is clear from context that ring homomorphism (epimorphism, etc.) is intended, we shall refer simply to "homomorphism," etc.

Note that by equation (5), η is a group homomorphism of the additive group of R.

By definition, every epimorphism is a homomorphism and every isomorphism is a monomorphism; in each case the converse is false.

EXAMPLE 9. Let R be a ring, and let T be a ring with $0'$ as zero element. Then $\eta : R \longrightarrow T$, defined by $\eta(x) = 0'$, $x \in R$, is clearly a homomorphism (details?). Such homomorphisms are said to be "trivial" or "zero homomorphisms."

EXAMPLE 10. Let n be a nonnegative integer and let $\nu_n : Z \longrightarrow Z_n$ be the mapping defined by

$$\nu_n(x) = \bar{x}, \quad x \in Z,$$

where $\bar{x} \in Z_n$ is the residue class mod n containing the integer x. Then

$$\nu_n(x + y) = \overline{x + y} = \bar{x} \oplus \bar{y} = \nu_n(x) \oplus \nu_n(y),$$

and

$$\nu_n(xy) = \overline{xy} = \bar{x} \odot \bar{y} = \nu_n(x) \odot \nu_n(y),$$

so that ν_n is a ring homomorphism. Since ν_n is a surjection (why?), it is an epimorphism.

EXAMPLE 11. Suppose R is a subring of a ring T. Then the inclusion mapping $i : R \longrightarrow T$ defined by

$$i(x) = x, \quad x \in R,$$

is a monomorphism. If $R = T$, then i is the identity mapping of T, whereas if $R \subsetneq T$, then i is the restriction of the identity mapping to R. In the latter case, i is a monomorphism but not an isomorphism.

The first theorem on ring homomorphisms is an analogue of Theorem 5, Chapter 3, 128.

THEOREM 20. Let R be a ring (not necessarily commutative) with binary operations $+$ and $\cdot$ and let $T \neq \emptyset$ be a set on which are defined two binary operations $\oplus$ and $\odot$. Further, let $\eta : R \longrightarrow T$ be a homomorphism and suppose that $\oplus'$ and $\odot'$ are the restrictions of $\oplus$ and $\odot$ to $R' = \eta[R]$. Then

(i) R' is a ring with respect to $\oplus'$ and $\odot'$, and $0' = \eta(0)$ is its zero element (in short, an epimorphic image of a ring is a ring);

(ii) for each $x \in R$, $\eta(-x) = -\eta(x)$;

(iii) if R has a unity element e, then $e' = \eta(e)$ is a unity element of R';

(iv) if R is commutative, so is R'.

> *Proof:* As compared with the proof of Theorem 5, Chapter 3, the only new feature is the distributivity of $\odot'$ over $\oplus'$. The proof is therefore left to the reader.

COROLLARY. Let R, T, η and R' be as in Theorem 20. If T is a ring, then R' is a subring of T.

> *Proof:* Exercise.

Exercises

1. Prove that an isomorphic image of an integral domain (alternatively, a field) is an integral domain (a field).

2. Prove: A composite of homomorphisms (alternatively, epimorphisms, monomorphisms, isomorphisms) is a homomorphism (epimorphism, monomorphism, isomorphism).

3. (i) Define *endomorphism* and *automorphism* of a ring. (ii) Show that a composite of automorphisms of a ring is an automorphism of the given ring. (iii) Prove that the set $\mathscr{A}(R)$ of all automorphisms of a ring R is a group with respect to composition of mappings.

4. (i) Exhibit a ring endomorphism $\eta : R \longrightarrow R$ which is a surjection but not an automorphism. (*Hint:* Use Example 8, Chapter 3, page 139, to construct a ring S.) (ii) Exhibit a ring monomorphism $\mu : R \longrightarrow R$ which is not an automorphism.

5. Let R be a subset of a ring T, and let
$$H = \{\alpha | \alpha \in \mathscr{A}(T) \text{ and } \alpha(x) = x \text{ for each } x \in R\}.$$
Prove that H is a subgroup of $\mathscr{A}(T)$. Set
$$K = \{\beta | \beta \in \mathscr{A}(T) \text{ and } \beta[R] = R\}.$$
Show that K is a subgroup of $\mathscr{A}(T)$ and that $H < K$.

6. Let T be a ring and let $G < \mathscr{A}(T)$. Prove that
$$R_G = \{x | x \in T \text{ and } \alpha(x) = x \text{ for all } \alpha \in G\}$$
is a subring of T.

7. Let R be a subring of a ring T and define H as in Exercise 5. Also define R_H as in Exercise 6. Prove that $R \subset R_H$.

8. Is it true that an epimorphic image of an integral domain is an integral domain? If not, give a counter example.

With Theorem 20 out of the way *we return to the universal assumption that all rings under consideration are commutative.* Let R and T be rings and let $\eta : R \longrightarrow T$ be a homomorphism. As observed earlier, η is a homomorphism of the additive group of R and (Theorem 20) $\eta(0) = 0'$ is the neutral element of the additive group of T. Since a ring homomorphism is a group homomorphism with an additional feature, it is reasonable to make

DEFINITION 13. The *kernel* of η is the set

$$\ker(\eta) = \{x \,|\, x \in R \text{ and } \eta(x) = 0'\}.$$

Since $\eta(0) = 0', 0 \in \ker (\eta)$ and therefore the kernel of a homomorphism is never empty.

By definition, the kernel of a ring homomorphism η is precisely the kernel of η considered as a homomorphism of the additive group of R. This being the case, it follows that if $a, b \in \ker (\eta)$, then $a - b \in \ker (\eta)$. Also, if $a, b \in \ker (\eta)$, then $\eta(ab) = \eta(a)\eta(b) = 0' \cdot 0' = 0'$, and so $ab \in \ker (\eta)$. Consequently, $\ker (\eta)$ is a subring of R (Theorem 19). However, the feature that distinguishes the kernels of homomorphisms from other subrings is this:

If $a \in \ker (\eta)$ and if $x \in R$, then $xa \in \ker (\eta)$.

For,

$$\eta(xa) = \eta(x)\eta(a) = \eta(x) \cdot 0' = 0'.$$

Consequently, $\ker (\eta)$ is not merely closed under multiplication, it is, so to speak, "superclosed."

DEFINITION 14. A nonempty subset I of a ring R is an *ideal* in R if and only if

(i) for all $a, b \in I$, $a - b \in I$; and
(ii) for each $x \in R$ and for each $a \in I$, $xa \in I$.

We have proved

THEOREM 21. The kernel of a ring homomorphism $\eta : R \longrightarrow T$ is an ideal in R.

Thus every homomorphism of a ring determines an ideal in the ring. The converse, that each ideal in a ring determines a homomorphism, will be proved in due course (see Theorem 30, page 178). And these two results will show that ideals in commutative rings occupy roles comparable to those of normal subgroups in group theory. In addition to being essential for the study of ring homomorphisms, we shall find (see especially Sections 12 and 13) that ideals are closely related to the concepts of divisibility and primeness of elementary arithmetic.

EXAMPLE 12. The subset $\{0\}$ of the ring R is an ideal in R, the *zero* ideal. The ring R, itself, is an ideal in R and is known as the "unit ideal." The term "unit ideal" has nothing whatsoever to do with the presence of a unity element. There are many rings which do not possess a unity element, but every ring has a unit ideal. The zero ideal and the unit ideal are the *trivial* ideals; all others are *nontrivial* or *proper*.

EXAMPLE 13.   The kernel of the homomorphism $\nu_n : Z \longrightarrow Z_n, n \in N$, defined in Example 10, page 165, is the set nZ, i.e.,

$$\ker (\nu_n) = nZ.$$

For, if $x \in Z$, then $\nu_n(x) = \bar{x} = x + nZ$. But $\bar{x} = \bar{0}$ if and only if $x = x - 0 \in nZ$; hence $\ker (\nu_n) = nZ$. By Theorem 21, nZ is an ideal in Z.

One can also verify that $nZ, n \in N$, is an ideal in Z by a direct application of Definition 14. Of course, the set nZ, where n is a negative integer, is an ideal in Z. But since $nZ = (-n)Z$, nothing is gained by permitting n to assume negative integral values.

EXAMPLE 14.   Lest the reader acquire the mistaken notion that every subring is an ideal, we now give a counterexample. Let $G = \{a + bi | a, b \in Z$ and $i^2 = -1\}$ be the set of Gaussian integers. With respect to the usual addition and multiplication of complex numbers, G is a ring. The ring of integers Z $(= \{a + bi | a \in Z$ and $b = 0\})$ is readily seen to be a subring of G. But if we take $x = i (\in G)$ and $a = 1 (\in Z)$, then

$$xa = i \cdot 1 = i \notin Z.$$

Thus condition (ii) of Definition 14 is violated and Z is not an ideal in G.

Exercises

1. Let $R = Z \times Z$ and define $\oplus$ and $\odot$ on R by $(a, b) \oplus (c, d) = (a + c, b + d)$ and $(a, b) \odot (c, d) = (ac, bd)$, respectively. Prove, in two ways, that $\{(0, y) | y \in Z\}$ is an ideal in R.

2. Prove that the integers (rationals) comprise a subring but not an ideal in the rationals (reals).

3. Let η be a homomorphism of a ring R. Prove: η is trivial if and only if $\ker (\eta) = R$; η is a monomorphism if and only if $\ker (\eta) = \{0\}$.

4. Suppose that $\xi : R \longrightarrow T$ is an epimorphism and $\eta : T \longrightarrow U$ is a homomorphism. Prove that $\ker (\xi) \subset \ker (\eta \circ \xi)$. Also, show that $\ker (\xi) = \ker (\eta \circ \xi)$ if and only if η is a monomorphism.

9.   *Ideals*

In examining the modular arithmetics we find that Z_4, Z_6, Z_8 and Z_9 have proper ideals whereas Z_2, Z_3, Z_5 and Z_7 do not. Since the latter are all fields but the former are not, our observation suggests

THEOREM 22.   A field contains no proper ideals.

Proof: Let us begin by showing that if R is a ring with unity element e, if I is an ideal in R, then $I = R$ if and only if $e \in I$.

Let $e \in I$. Since $I \subset R$ is automatic, we have to verify the reverse inclusion. Therefore let $x \in R$; since $e \in I$, by Definition 14, part (ii), $x = xe \in I$ and so $R \subset I$. Consequently, $R = I$. The converse is trivial.

Now suppose F is a field and I is an ideal in F. We shall prove that if $I \neq \{0\}$, then $I = F$; this shows that F has no proper ideals. Since F is a field, it has a unity element e. Further, if $I \neq \{0\}$, there is an element $a \in I$, $a \neq 0$. But $a \neq 0$ and $a \in F$ imply that a has a reciprocal a^{-1} in F. Therefore (by (ii), Definition 14) $e = a^{-1}a \in I$, and so by the initial remarks $I = F$.

Q.E.D.

COROLLARY.   A homomorphism of a field is either trivial or else it is a monomorphism.

Proof: Exercise.

Whatever kinds of ideals a ring may have, the following theorem is always true.

THEOREM 23.   Let $\mathscr{I}$ be a set of ideals in a ring R. Then $\bigcap \mathscr{I}$ is an ideal in R.

Proof: First of all, $\bigcap \mathscr{I} \neq \emptyset$ (why?). Next, suppose $a, b \in \bigcap \mathscr{I}$; then for each $I \in \mathscr{I}$, $a, b \in I$, and therefore $a - b \in I$. Hence $a - b \in \bigcap \mathscr{I}$ and this proves (i) of Definition 14. The reader should verify part (ii) of Definition 14 and thus complete the proof of the theorem.

Q.E.D.

Again, with $\mathscr{I}$ a set of ideals in R, it is natural to ask: "is $\bigcup \mathscr{I}$ an ideal in R?" In general, the answer is "no," and the reader should provide counterexamples. However, we do have a useful analogue of Theorem 23, namely,

THEOREM 24.   If $\mathscr{I} = \{I_j \mid j \in N\}$ is a set of ideals in R such that for each j, $I_j \subset I_{j+1}$, then $\bigcup \mathscr{I}$ is an ideal in R.

Proof: Exercise.

In the remainder of this section we consider some simple ideals in two different kinds of rings.

A. *Rings With Unity Element.* Let e be the unity element of the ring R and let $a \in R$. We claim that

(7) $$Ra = \{xa \mid x \in R\} \qquad (= aR)$$

is an ideal in R. For, if $u, v \in Ra$, then there are elements $x, y \in R$ such that $u = xa$ and $v = ya$. Therefore $u - v = (x - y)a$ where $x - y \in R$, and so $u - v \in Ra$. Further, if $t \in R$ and $xa \in Ra$, then

$$t(xa) = (tx)a$$

where $tx \in R$. Hence $t(xa) \in Ra$, and consequently Ra is an ideal.

Now since R has a unity element e, a is an element in the ideal Ra. For,

(8) $$a = ea \in Ra.$$

Moreover, if I is an ideal in R and $a \in I$, then for each $x \in R$, $xa \in I$. Consequently

$$Ra = \{xa \,|\, x \in R\} \subset I,$$

and so we see that the ideal Ra is a subset of every ideal containing the element a. Hence if we put

$$\mathscr{I} = \{I \,|\, I \text{ is an ideal in } R \text{ and } a \in I\},$$

then

$$Ra \subset \bigcap \mathscr{I}.$$

On the other hand, Ra is also an ideal containing a (see (8)), hence $Ra \in \mathscr{I}$. Consequently $\bigcap \mathscr{I} \subset Ra$, and we have proved

THEOREM 25. *If $a \in R$ and R is a ring with unity element, then*

$$Ra = \bigcap \mathscr{I}$$

where $\mathscr{I} = \{I \,|\, I \text{ is an ideal in } R \text{ and } a \in I\}$.

In essence, Theorem 25 states that in a ring R with unity element, Ra is the smallest ideal containing a.

DEFINITION 15. Let R be a ring, **WITH OR WITHOUT A UNITY ELE-MENT**, and let $a \in R$. The *principal ideal in R generated by a* is the ideal

$$\bigcap \mathscr{I}$$

where $\mathscr{I} = \{I \,|\, I \text{ is an ideal in } R \text{ and } a \in I\}$. We denote the principal ideal generated by a by

$$(a),$$

and a is the *generator* of (a).

COROLLARY. *If R is a ring with unity element and $a \in R$, then $Ra = (a)$.*

Exercises

1. Prove that in a field every ideal is principal.
2. May a principal ideal have more than one generator? Explain.

3. Let a, b be elements in a ring R with unity element. Verify that

$$Ra + Rb = \{xa + yb \,|\, x, y \in R\}$$

is an ideal in R and show that $\{a, b\} \subset Ra + Rb$. $Ra + Rb$ is *the ideal generated by a and b*, or *generated by the set* $\{a, b\}$. Prove that if $b = \pm a$, then $Ra + Rb = Ra(=Rb)$.

4. Prove that $Ra + Rb$ is a subset of every ideal containing a and b.

5. Let R be a ring with unity element, and let $\mathscr{I} = \{I \,|\, I$ is an ideal in R and $\{a, b\} \subset I\}$. Prove that $\cap \mathscr{I} = Ra + Rb$.

6. Let $R = Z \times Z$ where sum and product are defined by $(a, b) + (c, d) = (a + c, b + d)$ and $(a, b) \cdot (c, d) = (ac, bd)$, respectively. Note that $(1, 1)$ is the unity element of R. Describe, in a simple fashion, the ideal generated by the elements $(2, 0)$ and $(0, 3)$. Is this a principal ideal? Why?

7. Suppose that $a_1, a_2, \cdots, a_n$ are elements in the ring R with unity element e. Prove that

$$Ra_1 + Ra_2 + \cdots + Ra_n = \{x_1 a_1 + x_2 a_2 + \cdots + x_n a_n \,|\, x_i \in R, 1 \leq i \leq n\}$$

is an ideal in R containing the subset $\{a_1, a_2, \cdots, a_n\}$. $Ra_1 + Ra_2 + \cdots + Ra_n$ is *the ideal generated by the subset* $\{a_1, a_2, \cdots, a_n\}$, and $\{a_1, a_2, \cdots, a_n\}$ is a *set of generators* of the ideal.

8. Give an example of an ideal generated by a set of n ($n \geq 3$) elements, but not generated by a set containing fewer than n elements.

9. Prove that if I is an ideal in R containing the elements $a_1, a_2, \cdots, a_n$, then

$$Ra_1 + Ra_2 + \cdots + Ra_n \subset I.$$

Letting $\mathscr{I} = \{I \,|\, I$ is an ideal in R and $\{a_1, a_2, \cdots, a_n\} \subset I\}$, show that if R has a unity element, then

$$Ra_1 + Ra_2 + \cdots + Ra_n = \cap \mathscr{I}.$$

B. *Rings Without Unity Element.* Let R be a ring without unity element and let $a \in R$. Even though R contains no unity element we may, nevertheless, define Ra ($= aR$) by equation (7). Without modification, the argument on pages 169–170 shows that Ra is an ideal in R. However, now it may happen that a is *not* in the ideal Ra.

EXAMPLE 15. Let $R = 2Z$, the ring of even integers, and let $a = 4$. The ideal $Ra = (2Z)4$ does not contain the integer 4.

We shall still accept Definition 15 as the definition of *principal ideal*, but Example 15 shows that if R has no unity element, then Ra may *not* be the principal ideal generated by a. Now if $Ra \neq (a)$, how can we describe the elements of the latter ideal? Certainly $a \in (a)$, since a is an element in each ideal $I \in \mathscr{I}$ and $(a) = \cap \mathscr{I}$. Further, since (a) is a group with respect to $+$, we know that $a + a \in (a)$, $a + a + a \in (a)$, and in fact $na \in (a)$ for

each $n \in Z$. (Recall: If n is a positive integer, $na = \overbrace{a + a + \cdots + a}^{n \text{ times}}$; if $n = 0, 0a = 0$; if n is a negative integer, $na = \overbrace{(-a) + (-a) + \cdots + (-a)}^{-n \text{ times}}$.)
Suppose $x \in R$; by (ii) of Definition 14, $xa \in (a)$. And now since $na \in (a)$ and $xa \in (a)$,

$$na + xa \in (a), n \in Z, x \in R,$$

i.e.,

(9) $\{na + xa \mid n \in Z \text{ and } x \in R\} \subset (a).$

It must be emphasized that $na + xa$ cannot be factored to obtain $(n + x)a$ (why?). We claim that the inclusion (9) can be strengthened to equality.

For convenience, put $A = \{na + xa \mid n \in Z \text{ and } x \in R\}$. By Definition 15, (a) is a subset of every ideal containing the element a. Therefore our contention will be proved by verifying that $a \in A$ and that A is an ideal. Since $a = 1a + 0 \cdot a$, $0 \in R$, we have $a \in A$. All that's left is to prove A is an ideal. Let

$$n_i a + x_i a \in A, i = 1, 2;$$

then

$$(n_1 a + x_1 a) - (n_2 a + x_2 a) = (n_1 a - n_2 a) + (x_1 a - x_2 a).$$

But for all integers n_1 and n_2,

$$n_1 a - n_2 a = (n_1 - n_2)a = na,$$

where $n = n_1 - n_2 \in Z$ (why?), and by the distributive law in R, $x_1 a - x_2 a = (x_1 - x_2)a = xa$, where $x = x_1 - x_2 \in R$. Therefore

$$(n_1 a + x_1 a) - (n_2 a + x_2 a) = na + xa,$$

where $n \in Z$ and $x \in R$, so $(n_1 a + x_1 a) - (n_2 a + x_2 a) \in A$. Thus condition (i) of Definition 14 is satisfied.

Next let $y \in R$ and $na + xa \in A$; then

$$y(na + xa) = y(na) + y(xa).$$

But (assuming $n > 0$)

$$\begin{aligned}
y(na) &= y(\overbrace{a + a + \cdots + a}^{n \text{ times}}) \\
&= ya + ya + \cdots + ya \quad \text{(distributive law)} \\
&= (y + y + \cdots + y)a \\
&= (ny)a \quad\quad\quad\quad\ \text{(definition of } ny\text{)}
\end{aligned}$$

where $ny \in R$. Corresponding results hold for $n \leq 0$. Also, since $y, x, a \in R$, $r(xa) = (rx)a$. Hence

$$y(na + xa) = (ny)a + (yx)a$$
$$= (ny + yx)a$$
$$= 0a + (ny + yx)a,$$

where 0 is the *integer* 0 and $ny + yx \in R$. Therefore $y(na + xa) \in A$, and so (ii) holds. Consequently, A is an ideal in R. We have proved

THEOREM 26. If a is an element in the ring R (with or without unity element), then

$$(a) = \{na + xa \,|\, n \in Z \text{ and } x \in R\}.$$

We have seen that if R has a unity element, then (i) $a \in Ra$ and (ii) $Ra = (a)$. These results suggest

THEOREM 27. Let a be an element in a ring R, with or without a unity element. Then $Ra = (a)$ if and only if $a \in Ra$.

Proof: If $a \in Ra$, then since Ra is an additive Abelian group, $na \in Ra$ for each $n \in Z$, and also $xa \in Ra$ for each $x \in R$. Therefore

$$(a) = \{na + xa \,|\, n \in Z \text{ and } x \in R\} \subset Ra.$$

Since the reverse inclusion holds universally, the first half of the theorem is proved.

Conversely, suppose $Ra = (a)$. Since $a \in (a)$, we have $a \in Ra$ and the proof is complete.

Q.E.D.

For the most part we shall study rings with unity element; consequently, in every case the principal ideal generated by the element a is

$$Ra = aR.$$

Exercises

1. Let $R = Z \times 2Z$. Verify that (i) if $a = (1, 0)$, then $Ra = (a)$; (ii) if $a = (0, 2)$, then

$$Ra \subsetneq (a).$$

2. Let $R = Z \times Z \times 2Z$ and define sum and product of elements in R by $(x, y, z) + (x', y', z') = (x + x', y + y', z + z')$ and $(x, y, z) \cdot (x', y', z') = (xx', yy', zz')$, respectively. Set $a = (1, 0, 0)$, $b = (0, 1, 0)$ and show that (i) $Ra + Rb = \{ma + nb + ua + vb \,|\, m, n \in Z \text{ and } u, v \in R\}$; (ii) show that if $a = (1, 0, 0)$ and $b = (0, 0, 2)$, then $Ra + Rb \subsetneq \{ma + nb + ua + vb \,|\, m, n \in Z \text{ and } u, v \in R\}$.

3. Let a, b be elements in a ring R, and put $\mathscr{I} = \{I \,|\, I \text{ is an ideal in } R \text{ and } \{a, b\} \subset I\}$. Prove (i) $\bigcap \mathscr{I} = (a, b)$ is an ideal containing the elements a and b, and (ii) $(a, b) = \{ma + nb + xa + yb \,|\, m, n \in Z \text{ and } x, y \in R\}$. Prove that (iii) $Ra + Rb \subset (a, b)$; (iv) $Ra + Rb = (a, b)$ if and only if $\{a, b\} \subset Ra + Rb$; and (v) if R has a unity element, then $Ra + Rb = (a, b)$.

DEFINITION 16. (a, b) is the *ideal generated by the elements* a and b; a and b are *generators* of (a, b).

4. (Generalization of Exercise 3) If $a_1, a_2, \cdots, a_r, r \geq 1$, are elements in a ring R, let $\mathscr{I} = \{I \mid I$ is an ideal in R and $\{a_1, a_2, \cdots, a_r\} \subset I\}$. Prove (i) $\bigcap \mathscr{I} = (a_1, a_2, \cdots, a_r)$ is an ideal in R and $\{a_1, a_2, \cdots, a_r\} \subset (a_1, a_2, \cdots, a_r)$; (ii) $Ra_1 + Ra_2 + \cdots + Ra_r \subset (a_1, a_2, \cdots, a_r)$; (iii) $Ra_1 + Ra_2 + \cdots + Ra_r = (a_1, a_2, \cdots, a_r)$ if and only if $\{a_1, a_2, \cdots, a_r\} \subset Ra_1 + Ra_2 + \cdots + Ra_r$; (iv) if R has a unity element, then $Ra_1 + Ra_2 + \cdots + Ra_r = (a_1, a_2, \cdots, a_r)$; (v) $(a_1, a_2, \cdots, a_r) = \{n_1a_1 + n_2a_2 + \cdots + n_ra_r + x_1a_1 + x_2a_2 + \cdots + x_ra_r \mid n_i \in Z$ and $x_i \in R$, $1 \leq i \leq r\}$.

DEFINITION 16′. The ideal $(a_1, a_2, \cdots, a_r)$ is the *ideal generated by the set* $\{a_1, a_2, \cdots, a_r\}$ and $\{a_1, a_2, \cdots, a_r\}$ is a *set of generators* of $(a_1, a_2, \cdots, a_r)$.

5. (Generalization of Exercise 4) Let S be a subset of a ring R, and let $\mathscr{I} = \{I \mid I$ is an ideal in R and $S \subset I\}$. Prove: (i) $\bigcap \mathscr{I} = (S)$ is an ideal in R containing the subset S; (ii) $(S) = \{u \mid$ there is an integer $r \geq 0$, there exist integers $n_1, n_2, \cdots, n_r$, there exist elements $a_1, a_2, \cdots, a_r$ in S, and there exist elements $x_1, x_2, \cdots, x_r$ in R such that $u = n_1a_1 + n_2a_2 + \cdots + n_ra_r + x_1a_1 + x_2a_2 + \cdots + x_ra_r\}$; (iii) if R has a unity element, then $(S) = \{u \mid$ there is an integer $r \geq 0$, there exist elements $a_1, a_2, \cdots, a_r$ in S, and there exist elements $x_1, x_2, \cdots, x_r$ in R, such that $u = x_1a_1 + x_2a_2 + \cdots + x_ra_r\}$.

DEFINITION 16″. (S) is the *ideal generated by the set* S and S is a *set of generators* for (S).

6. Let I be an ideal in a ring R with unity element, let $x \in R$, and let A be the ideal in R generated by the set $I \cup \{x\}$. Prove: $y \in A$ if and only if there is an element $u \in I$ and an element $v \in R$ such that $y = u + vx$. Show that $I \subsetneqq A$ if and only if $x \notin I$.

7. Prove: If I is an ideal in Z, then there is an integer $n \geq 0$ such that $I = nZ$. (*Hint:* Use the fact that an ideal is first and foremost a subgroup. What are the subgroups of Z?) In view of Example 13, page 168, we deduce: I is an ideal in Z if and only if there is an integer $n \geq 0$ such that $I = nZ$.

10. *Residue Class Rings*

In constructing the modular arithmetics from the domain of integers (Section 4), we followed the pattern for constructing a quotient group from a given group G and a normal subgroup of G. This method worked because every subgroup of the additive group of integers is, in fact, an ideal in the ring of integers (Example 13, page 168). This fact is used, in a disguised form, to prove Theorem 24, Chapter 2, page 98. And this theorem is essential for the purpose of defining a product of residue classes of integers. To clarify the situation, recall that the theorem states:

If a and b are integers such that $a \equiv b \pmod{n}$, then for each integer x, $ax \equiv bx \pmod{n}$ and $xa \equiv xb \pmod{n}$.

We now give an alternate proof using, explicitly, the fact that each subgroup nZ of Z $(n \geq 0)$ is an ideal in Z.

We know that $a \equiv b \pmod{n}$ if and only if $a - b \in nZ$. Since nZ is an ideal in Z, for each $x \in Z$, $ax - bx = x(a - b) \in nZ$. Hence $ax \equiv bx \pmod{n}$.

Next consider the problem: Given a ring R and an ideal I in R, can we mimic the construction of the modular arithmetics so as to produce a new ring from the residue classes of R with respect to I? Not only is the answer "yes," but we shall see that the new procedure differs only in terminology from the earlier one.

If I is an ideal in a ring R, then with respect to addition, I is a subgroup of the additive group of R. By Exercise 5, Chapter 2, page 116, the quotient group R/I is an Abelian group with respect to $\oplus$ where, if $A, B \in R/I$, and if $a \in A$, $b \in B$, then $A \oplus B$ is the coset containing the element $a + b$. Our task is to define a multiplication $\odot$ on R/I so that R/I, together with $\oplus$ and $\odot$, is a ring.

Since I is a subgroup of the additive (Abelian) group of R, I defines a stable equivalence relation $\sim$ on R. By analogy with the modular arithmetics, this equivalence relation is denoted by

$$\equiv \pmod{I};$$

if $a \sim b$, we write

$$a \equiv b \pmod{I},$$

i.e., *a is congruent to b modulo the ideal I*. From the definition of $\equiv \pmod{I}$ it is clear that:

1. for all $a, b \in R$, $a \equiv b \pmod{I}$ if and only if $a - b \in I$; therefore $a \equiv 0 \pmod{I}$ if and only if $a \in I$;
2. for each $a \in R$, $a \equiv a \pmod{I}$;
3. if $a, b \in R$ and $a \equiv b \pmod{I}$, then $b \equiv a \pmod{I}$;
4. if $a, b, c \in R$ and $a \equiv b \pmod{I}$ and $b \equiv c \pmod{I}$, then $a \equiv c \pmod{I}$.

The fact that $\equiv \pmod{I}$ is stable with respect to the group operation $+$ is expressed by: If $a \equiv b \pmod{I}$, then $a + x \equiv b + x \pmod{I}$ and $x + a \equiv x + b \pmod{I}$. In addition to stability relative to $+$, the following analogue of Theorem 24, Chapter 2, page 98, is valid.

THEOREM 28. If $a \equiv b$ (mod I), then for each $x \in R$, $ax \equiv bx$ (mod I) and $xa \equiv xb$ (mod I).

> *Proof:* Aside from a few changes of notation, the proof of Theorem 24, Chapter 2, carries over to the present theorem.
>
> Q.E.D.

COROLLARY. If $a \equiv b$ (mod I) and $c \equiv d$ (mod I), then $ac \equiv bd$ (mod I).

> *Proof:* Exercise.

Now let $A, B \in R/I$ and let $a \in A$, $b \in B$. Since $ab \in R$ and since R/I is a partition of R, there is a unique $C \in R/I$ such that $ab \in C$. Define

$$A \odot B = C;$$

to prove that $\odot$ is a binary operation on R/I, it suffices to show that if $x \in A$ and $y \in B$, then also $xy \in C$. From this result it follows that $\odot$ is independent of the choice of elements in A and B, respectively, therefore that $A \odot B$ depends only on the cosets A and B. But $x, a \in A$ if and only if $x - a \in I$, hence if and only if $x \equiv a$ (mod I). Similarly, $y, b \in B$ if and only if $y \equiv b$ (mod I). Therefore if $x, a \in A$ and $y, b \in B$, then $xy \equiv ab$ (mod I). Since $ab \in C$, also $xy \in C$, and therefore $\odot$ is a binary operation on R/I.

THEOREM 29. R/I is a commutative ring with respect to the binary operations $\oplus$ and $\odot$. If R has a unity element, so does R/I.

> *Proof:* Since R/I is an Abelian group with respect to $\oplus$, one completes the proof that R/I is a commutative ring by showing that $\odot$ is commutative and that (2) and (3) of Definition 1, page 146, are satisfied. The proofs of these statements are essentially carbon copies of the corresponding parts of Theorem 9, page 152, and are left to the reader.
>
> Now suppose e is the unity element of R and let $\bar{e}$ be the element in R/I such that $e \in \bar{e}$. We claim that $\bar{e}$ is the unity element of R/I. For, if $A \in R/I$, then $\bar{e} \odot A$ is the element of R/I containing ea where $a \in A$. Since $ea = a$, it follows that $\bar{e} \odot A = A$.
>
> Q.E.D.

DEFINITION 17. The ring R/I with binary operations $\oplus$ and $\odot$ as defined above is the *residue class ring of R modulo* (or, *with respect to*) *the ideal I*. The elements of R/I are the *residue classes* of I (or, *modulo I*).

If A is a residue class with respect to I, then A is also a coset of I considered as a subgroup of the additive group of R. Hence there is an $a \in A$ such that $A = a + I$. The residue class $A = a + I$ is henceforth denoted by "$\bar{a}$";

$$\bar{a} = a + I.$$

Exercises

Prove: For each $\bar{a}$ and each $\bar{b}$ in R/I,

1. $\bar{a} = \bar{b}$ if and only if $a - b \in I$;
2. $\bar{a} \oplus \bar{b} = (a + b) + I = \overline{a + b}$;
3. $\bar{a} \odot \bar{b} = ab + I = \overline{ab}$;
4. if e is a unity element in R and I is an ideal in R, then $\bar{e} = \bar{0}$ if and only if $I = R$.

So far, the only residue class rings we have encountered are the modular arithmetics. Of course, there are many others—for each commutative ring R and for each ideal I in R, there is a residue class ring R/I. To be sure, each of these might be isomorphic to some Z_n. For the present, we offer one example of a residue class ring which is *not* isomorphic to a modular arithmetic.

EXAMPLE 16.   Let G be the domain of Gaussian integers, and let $3G = \{3\alpha \mid \alpha \in G\}$ be the principal ideal generated by 3. We shall prove that $G/3G$ is a field containing exactly nine elements.

If $\alpha \in G$, let $\bar{\alpha} = \alpha + 3G$ be the residue class containing α. First, the nine residue classes,

(11) $\overline{x + yi}$, $x, y = 0, 1$, or 2,

are distinct. Indeed, suppose $\overline{a + bi} = \overline{c + di}$; then $(a + bi) - (c + di) = (a - c) + (b - d)i \in 3G$, hence 3 is a factor of $(a - c) + (b - d)i$. By the properties of the complex numbers,

(12) $3 \mid a - c$ and $3 \mid b - d$.

But if $a + bi$ and $c + di$ are among the elements (11), then $-3 < a - c < 3$ and $-3 < b - d < 3$. Therefore (12) implies that $a = c$ and $b = d$. Hence the residue classes (11) are distinct.

Next, if $\bar{\alpha} \in G/3G$, then $\bar{\alpha}$ is one of the residue classes (11). For, let $\alpha = u + vi$; by the division algorithm for Z there exist unique pairs of integers q, r and s, t such that

$$u = 3q + r   \quad \text{and} \quad   v = 3s + t$$

where $0 \le r < 3$ and $0 \le t < 3$. Then

$$u + vi = 3(q + si) + (r + ti).$$

Since $3(q + si) \in 3G$,

$$\overline{u + vi} = (r + ti) + 3G = \overline{r + ti},$$

which shows that $\overline{u + vi} = \bar{\alpha}$ is one of the residue classes (11).

Further, $G/3G$ contains no divisors of zero. Suppose $\overline{a + bi}$ and $\overline{c + di}$ are elements (11) such that

$$\overline{a + bi} \odot \overline{c + di} = \bar{0}$$

where $\overline{a + bi} \ne \bar{0}$. Since $\overline{a + bi} \ne \bar{0}$, $a \ne 0$ or $b \ne 0$. Now $\overline{a + bi} \odot \overline{c + di}$

$= \overline{(ac - bd) + (ad + bc)i}$, and this product is $\bar{0}$ if and only if $3|ac - bd$ and $3|ad + bc$. However, since the values of a, b, c, and d are 0, 1, or 2, and since $a \neq 0$ or $b \neq 0$, a simple computation shows that 3 is not a factor of both $ac - bd$ and $ad + bc$ unless $c = d = 0$. Hence $\overline{c + di} = \bar{0}$ and our assertion is proved.

We have proved that $G/3G$ is a finite integral domain. By Exercise 2, page 157, $G/3G$ is a field.

Finally, $G/3G$ is isomorphic to no Z_n. The only modular arithmetic having nine elements is Z_9. But since Z_9 is not a field, by Exercise 1, page 166, Z_9 and $G/3G$ are not isomorphic.

Exercise: Describe in detail the residue class ring $G/(1 + i)G$.

The question, "What are the conditions for a residue class ring R/I to be a field?" will be answered in Theorem 38, page 194.

11. *Some Basic Homomorphism Theorems*

In Theorem 21, page 167, we proved that every homomorphism of a ring determines an ideal in the ring, namely, the kernel of the homomorphism. The next result is a converse of Theorem 21. Its purpose is to show that given a ring R and an ideal I in R, there is a homomorphism of R whose kernel is precisely I.

THEOREM 30. Let R be a ring and I an ideal in R. Then there exists an epimorphism $\nu_I : R \longrightarrow R/I$ such that $\ker(\nu_I) = I$.

Proof: By analogy with Example 10 (page 165), define

$$\nu_I(x) = \bar{x}, \quad x \in R,$$

where $\bar{x}$ is the residue class modulo I containing the element x. Since each $x \in R$ is in some residue class modulo I, and since the set of residue classes partitions R, it follows that ν_I is a surjection. Moreover, if $x, y \in R$, then

$$\begin{aligned}
\nu_I(x + y) &= \overline{x + y} \\
&= \bar{x} \oplus \bar{y} \\
&= \nu_I(x) \oplus \nu_I(y),
\end{aligned}$$

and

$$\begin{aligned}
\nu_I(xy) &= \overline{xy} \\
&= \bar{x} \odot \bar{y} \\
&= \nu_I(x) \odot \nu_I(y),
\end{aligned}$$

whence ν_I is an epimorphism.

To prove that ker $(v_I) = I$, suppose first that $x \in$ ker (v_I). Then $v_I(x) = \bar{x}$ is the zero element of R/I, i.e., $\bar{x} = x + I = I$ and so $x \in I$. Hence ker $(v_I) \subset I$. Conversely, if $x \in I$, then $x + I = I$ and, by definition of v_I,

$$v_I(x) = I = \text{zero element of } R/I.$$

Therefore $x \in$ ker (v_I), whence $I \subset$ ker (v_I). Consequently ker (v_I) $= I$.

Q.E.D.

DEFINITION 18.   The epimorphism $v_I : R \longrightarrow R/I$ is the *quotient* (or *canonical*, or *natural*) epimorphism of R with respect to the ideal I.

As with group epimorphisms, we now have the following situation:

If $\eta : R \longrightarrow S$ is a ring epimorphism, then the kernel of η is an ideal in R (Theorem 21). By Theorem 30 there is a quotient epimorphism $v_{\text{ker}(\eta)} : R \longrightarrow R/\text{ker}(\eta)$, and the kernel of $v_{\text{ker}(\eta)}$ is also ker(η). Diagrammatically, these results are described by:

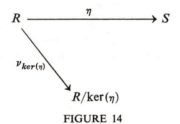

FIGURE 14

In view of Theorem 8, Chapter 3, page 132, it is natural to ask if there exists an isomorphism $\sigma : R/\text{ker}(\eta) \longrightarrow S$ closing the triangle of Figure 14, thus

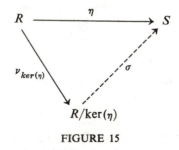

FIGURE 15

In other words, does there exist an isomorphism $\sigma : R/\ker(\eta) \longrightarrow S$ such that $\sigma \circ \nu_{\ker(\eta)} = \eta$? The proof of the existence and uniqueness of such an isomorphism parallels, in all essential details, the proof of Theorem 8, Chapter 3. We therefore state, without further ado,

THEOREM 31. If $\eta : R \longrightarrow S$ is an epimorphism, then there exists a unique isomorphism $\sigma : R/\ker(\eta) \longrightarrow S$ such that

$$(13) \qquad\qquad \sigma \circ \nu_{\ker(\eta)} = \eta.$$

Theorems 21, 30 and 31 show that ideals play a role in the theory of commutative rings, corresponding to that of normal subgroups in the theory of groups.

Exercises

1. Let $R = Z \times Z$ be the ring defined in Exercise 10, page 164, and let $S = Z$. Define

$$\eta((a, b)) = a, \quad (a, b) \in R.$$

(i) Prove that $\eta : R \longrightarrow S$ is an epimorphism. (ii) What is $\ker(\eta)$? (iii) Exhibit the isomorphism $\sigma : R/\ker(\eta) \longrightarrow S$ such that equation (13) holds.

2. State and prove analogues for rings of the First and Second Isomorphism Theorems for groups.

12. Principal Ideal and Unique Factorization Domains

Of all the integral domains the most familiar is the domain Z of integers. This domain has an interesting and highly developed arithmetic theory whose origins reach back to antiquity and whose development continues today with great vigor. In this section we consider the question: To what extent can the arithmetic of the integers be duplicated in an arbitrary integral domain? In seeking answers we shall see that the *principal ideal* and *unique factorization* properties play important roles.

DEFINITION 19. A ring R is a *principal ideal ring* if and only if each ideal in R is principal.

In the following we are concerned, almost exclusively, with principal ideal domains (PID), namely, integral domains which are principal ideal rings. The most familiar example of a PID is the domain Z of integers

(Exercise 7, page 174). Also by Exercise 1, page 170, every field is a PID. A further example is

EXAMPLE 17. The domain G of Gaussian integers is a PID. To prove this statement, it is convenient to regard G as a subdomain of the ring

$$G' = \{r + si \mid r, s \in R\},$$

where R is the field of rational numbers, and sum and product are defined for G' by

$$(r + si) + (u + vi) = (r + u) + (s + v)i$$

and

$$(r + si)(u + vi) = (ru - sv) + (rv + su)i,$$

respectively. Define a mapping $N: G' \longrightarrow R^+$ of G' into the set of nonnegative rationals by

$$N(r + si) = r^2 + s^2, \quad r + si \in G'.$$

Thus, if $\alpha = r + si \in G'$ and if $\alpha^* = r - si$ is the complex conjugate of α, $N(\alpha) = \alpha\alpha^*$. The definition of N (called a "norm") shows at once that $N(\alpha) \geq 0$ and the relationship $\alpha\alpha^* = N(\alpha)$ shows that $N(\alpha)$ is obtained by *rationalizing* the complex number α. $N(\alpha) = 0$ if and only if $\alpha = 0$. Clearly, if $\alpha, \beta \in G'$

(14) $$N(\alpha\beta) = N(\alpha)N(\beta);$$

and if $\alpha \in G$, then evidently $N(\alpha)$ is a nonnegative integer.

For those who are familiar with the Argand diagram for the complex numbers, it may be of interest to observe that the Gaussian integers are the lattice points of the complex plane and the elements of G' are represented by the rational points of the plane. The norm $N(\alpha)$ of a complex number $\alpha \in G'$ is the square of the distance of α from the origin.

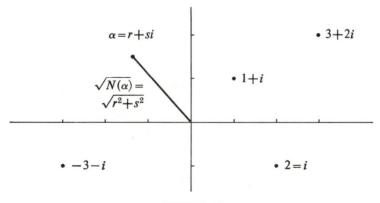

FIGURE 16

G' is a field; for, let $\alpha \in G'$, $\alpha \neq 0$. Then $\alpha^*/N(\alpha) \in G'$ and $\alpha \cdot \alpha^*/N(\alpha)$ $= 1$, hence $\alpha^*/N(\alpha)$ is the reciprocal of α.

Our proof that G is a PID requires two lemmas of which the first is a technical detail.

LEMMA 1. For each rational number r, there is an integer n such that $|n - r| \leq \frac{1}{2}$.

Proof: By the Archimedean property for the rationals, there is an integer k such that

$$k \leq r \text{ and } r < k + 1.$$

Now if $r - k > \frac{1}{2}$, then

$$1 = (k + 1) - k = ((k + 1) - r) + (r - k) > ((k + 1) - r) + \frac{1}{2},$$

whence

$$\frac{1}{2} > (k + 1) - r.$$

Therefore $r - k \leq \frac{1}{2}$ or $(k + 1) - r \leq \frac{1}{2}$. Since $|k - r| = r - k$ and $|(k + 1) - r| = (k + 1) - r$, the lemma follows.

$$\text{Q.E.D.}$$

LEMMA 2. If $\alpha, \beta \in G$ where $\beta \neq 0$, then there exist γ and δ in G such that

$$\alpha = \gamma\beta + \delta$$

where $N(\delta) < N(\beta)$.

Proof: Since G is a subset of the field G', there is a $\gamma' \in G'$ such that

$$(15) \qquad \alpha = \gamma'\beta \qquad \qquad \cdot$$

where $\gamma' = r + si$ and r, s are rational numbers. By Lemma 1, there exist integers n and m such that $|n - r| \leq \frac{1}{2}$ and $|m - s| \leq \frac{1}{2}$. Set $\gamma = n + mi$ and $\delta = \alpha - \gamma\beta$. Then

$$\alpha = \gamma\beta + \delta$$

where γ and δ are Gaussian integers; all that is left is to prove that $N(\delta) < N(\beta)$. Set $\eta = \gamma' - \gamma$; then

$$\begin{aligned}
\delta &= \alpha - \gamma\beta \\
&= (\gamma' - \gamma)\beta \qquad \text{(by equation (15))} \\
&= \eta\beta.
\end{aligned}$$

Hence by (14), $N(\delta) = N(\eta\beta) = N(\eta)N(\beta)$. But $\eta = (r - n) + (s - m)i$, and by definition of the norm,

$$\begin{aligned}
N(\eta) &= (r - n)^2 + (s - m)^2 \\
&= |n - r|^2 + |m - s|^2 \\
&\leq \left(\frac{1}{2}\right)^2 + \left(\frac{1}{2}\right)^2 = \frac{1}{2} < 1.
\end{aligned}$$

Therefore

$$N(\delta) = N(\eta)N(\beta) \leq \frac{1}{2} N(\beta) < N(\beta).$$

Q.E.D.

Finally,

THEOREM 32. *G* is a PID.

Proof: Let *I* be an ideal in *G*. If $I = \{0\}$, then $I = 0 \cdot G$ and *I* is principal. Suppose $I \neq \{0\}$; then there is a $\nu \in I$ such that $N(\nu) > 0$. Let

$$K = \{k \mid k \text{ is a positive integer and there is}$$
$$\text{a } \mu \in I \text{ such that } N(\mu) = k\}.$$

K is nonempty and, by the WOP, *K* contains a smallest positive integer *m*. Hence there is a nonzero $\beta \in I$ such that $N(\beta) \leq N(\alpha)$ for all nonzero α in *I*. Clearly, $\beta G \subset I$; to prove that *I* is a principal ideal, it suffices to verify the reverse inclusion. If $\alpha \in I$, then by Lemma 2 there exist γ and δ in *G* such that

$$\alpha = \gamma\beta + \delta$$

where $0 \leq N(\delta) < N(\beta)$. But $\alpha \in I$ and $\gamma\beta \in I$ imply that $\delta = \alpha - \gamma\beta \in I$. Since β is an element of *I* of smallest positive norm, and since $0 \leq N(\delta) < N(\beta)$, it follows that $N(\delta) = 0$, therefore $\delta = 0$. Hence $\alpha = \gamma\beta \in \beta G$ and so $I \subset \beta G$.

Q.E.D.

Exercise: Using the norm defined by $N(\alpha) = \alpha\alpha^*$, $\alpha \in G$, prove that $G/3G$ (Example 16) is a field containing nine elements. (*Hint:* First show that $\alpha \in 3G$ if and only if $3 \mid N(\alpha)$ and that $G/3G$ is a ring containing nine elements. Then prove that $G/3G$ is an integral domain by using the fact that if $3 \mid N(\alpha)N(\beta)$, then $3 \mid N(\alpha)$ or $3 \mid N(\beta)$.) Note that the elements of $G/3G$ have the form $x + y\gamma$ where $x, y \in Z_3$, and γ is the residue class of *j*. The field $G/3G$ is usually denoted by "$Z_3[\gamma]$."

Other important PID's are the *polynomial domains* to be studied in detail in Chapter 5. To give an example of an integral domain which is not a PID, we wait until some consequences of the principal ideal property have been deduced.

In developing an arithmetic of an integral domain comparable with that of the integers, we require a few simple concepts.

DEFINITION 20. An element *x* in an integral domain *D* is a *unit* in *D* if and only if there is a $y \in D$ such that $xy = e$. In this case, *y* is also a unit in *D*.

Since $ee = e$, it is clear that the unity element is a unit. The units in Z are 1 and -1; every nonzero element in a field is a unit; and in the Gaussian integers G, the only units are 1, -1, i, and $-i$ (details?). Those integral domains having no unity element also have no units, e.g., $2Z$.

DEFINITION 21.   Let x and y be elements in an integral domain D where $x \neq 0$. Then x is a *divisor* (or, *factor*) of y, $x \mid y$, if and only if there is a $z \in D$ such that $y = zx$; x is a *proper* divisor of y if neither x nor z is a unit in D. Further, if x and y are nonzero elements in D, then $d \in D$ is a *greatest common divisor* (g.c.d.) of x and y if and only if
  (i) $d \mid x$ and $d \mid y$;
  (ii) $f \mid x$ and $f \mid y$ imply $f \mid d$, where $f \in D$.
Finally, if x and y are nonzero elements in D, then x and y are *associates* if and only if $x \mid y$ and $y \mid x$.

The study of divisibility in a field F is trivial. Indeed, for each nonzero element $x \in F$ and for each $y \in F$, one always has $x \mid y$. On the other hand, in the domains Z and G such universal divisibility is not valid.

Exercises

1. Prove that a product of units in an integral domain is a unit.

2. Prove that for each integer n, the ring nZ is a principal ideal ring. (*Hint:* Use the WOP.)

3. Let D be an integral domain and let U be the set of units in D. Prove that U is a group with respect to (the restriction of) multiplication as group operation.

4. Does each pair of integers in the ring $2Z$ have a g.c.d.? Why?

5. The concept of g.c.d. can be defined for rings other than integral domains. Suppose D is a ring without divisors of zero. Prove: If each pair of nonzero elements of D possesses a g.c.d., then D has a unity element. (*Hint:* Prove and use the fact that if $az = z$ for some nonzero $z \in D$, then a is a unity element in D.)

6. Suppose that x and y are nonzero elements in an integral domain D. Prove that x and y are associates if and only if there exist units u and v in D such that $x = uy$ and $y = vx$. Define $x \sim y$ if and only if x and y are associates. Show that $\sim$ is an equivalence relation on D.

7. Let D be an integral domain. Prove that $Da \subset Db$ if and only if $b \mid a$. Further, $Da \subsetneqq Db$ if and only if b is a proper divisor of a.

THEOREM 33.   Each pair of nonzero elements a and b in a PID has a g.c.d. If d and d' are g.c.d.'s of a and b, then d and d' are associates.

Remark: Exercises 2 and 4 above show that Theorem 33 is not true for rings without zero divisors which also lack a unity element.

Proof: Set

(16) $$I = Ra + Rb;$$

I is an ideal in D (Exercise 3, page 171), and by hypothesis there is a $d \in D$ such that

(17) $$I = dD = \{dz \mid z \in D\}.$$

Since D has a unity element, a, b and d are all elements in I. Hence, by (17) there exist $p, q \in D$ such that $a = dp$ and $b = dq$, and so (i) of Definition 21 holds.

Suppose $d' \mid a$ and $d' \mid b$; we prove that $d' \mid d$. Since d' is a divisor of a and of b, there are $r, s \in D$ such that $a = d'r$ and $b = d's$. On the other hand, since $d \in I$, by (16) there are t and u in D such that

$$\begin{aligned} d &= ta + ub \\ &= td'r + ud's \\ &= (tr + us)d', \end{aligned}$$

where $tr + us \in D$. Thus $d' \mid d$ and so (ii) is proved. Hence d is a g.c.d. of a and b.

Now suppose both d and d' are g.c.d.'s of a and b. By (ii), Definition 21, $d \mid d'$ and $d' \mid d$ where $d' = xd$ and $d = yd'$ for some $x, y \in D$. Then $d = yxd$, and by the cancellation law for multiplication, $yx = e$. Therefore, x and y are units in D and the theorem is proved.

Q.E.D.

If d is a g.c.d. of a and b in a PID, one speaks of d as *the* g.c.d. of a and b, meaning thereby that the only other g.c.d.'s are associates of d. The g.c.d. of a and b is denoted by "g.c.d. (a, b)."

Exercises

1. Prove: If a, b are nonzero elements in a PID, then there are elements s, t in the domain such that $sa + tb = $ g.c.d. (a, b).

2. Let a and b be nonzero integers, and let $d = $ g.c.d. (a, b) where $a = dx$ and $b = dy$. Suppose s and t are integers such that $sa + tb = d$. Prove: (a) For each $k \in Z$, if $s' = s + ky$ and $t' = t - kx$, then $s'a + t'b = d$. (b) Conversely, if $s'a + t'b = d$, then there is an integer k such that $s' = s + ky$ and $t' = t - kx$.

3. Let S be a PID which is a subdomain of an integral domain R. Prove: If $d = $ g.c.d. (a, b) where $a, b, d \in S$, then d is the g.c.d. of a and b in R.

Since we are attempting to copy the arithmetic of the integers in other integral domains, let us make

DEFINITION 22.   Let D be an integral domain and let $p \in D$, where p is nonzero and a nonunit. Then p is *prime* if and only if $p = ab$ implies a or b is a unit; if p is not prime, it is *composite*.

In the domain of integers, 2, 3, 5, etc., are primes. Every nonzero element in a field is a unit, hence a field contains no primes or composite elements. The domain G of Gaussian integers contains primes. Indeed, if $\alpha \in G$ and if $N(\alpha) = a^2 + b^2$ (where $\alpha = a + bi$) is a prime in Z, then α is a prime in G. To see this, one proves first: if δ is nonzero and a nonunit, then $N(\delta) > 1$. Now suppose $\alpha = \beta\gamma$; then $N(\alpha) = N(\beta)N(\gamma)$. If neither β nor γ is a unit, then $N(\beta) > 1$ and $N(\gamma) > 1$, and so $N(\alpha)$ is not a prime, a contradiction. For instance, since $N(1 + i) = 2$ is a prime in Z, $1 + i$ is a prime in the Gaussian integers.

As the next example shows, a given element may be prime in one integral domain but composite in another.

EXAMPLE 18.   Let $Z[\sqrt{3}] = \{a + b\sqrt{3} \,|\, a, b \in Z\}$ where sum and product are defined by

$$(a + b\sqrt{3}) + (c + d\sqrt{3}) = (a + c) + (b + d)\sqrt{3},$$

and

$$(a + b\sqrt{3}) \cdot (c + d\sqrt{3}) = (ac + 3bd) + (ad + bc)\sqrt{3},$$

respectively. Clearly, $Z[\sqrt{3}]$ is an integral domain containing Z as a subdomain. In Z, 3 is a prime, whereas in $Z[\sqrt{3}]$, $\sqrt{3} \cdot \sqrt{3} = 3$. Since $\sqrt{3}$ is not a unit in $Z[\sqrt{3}]$ (why?), 3 is not a prime in $Z[\sqrt{3}]$.

In the arithmetic of the integers Z, a basic result is: Every integer different from 0, 1, and -1 is a product of powers of primes. Each such product is unique up to order of the factors, and to within units. The proof of this fundamental theorem rests heavily upon the fact that Z is a PID. It is therefore plausible to conjecture that the same result is valid in every PID. Lemmas 1, 2 and 3 and Theorems 34a and 34b verify this conjecture. *Throughout these lemmas and theorems, it is assumed that D is a PID and $a_0 \in D$ is neither zero nor a unit.* We emphasize that although Lemmas 1-3 serve the immediate purpose of preparing the proofs of Theorems 34a and 34b, they are intrinsically important results and are used extensively in the theory of integral domains.

LEMMA 1.   There exists no set $A = \{a_i \,|\, a_i \in D, \, i \in N$, and for each i, a_{i+1} is a proper divisor of $a_i\}$.

In other words, if $\{a_0, a_1, a_2, \cdots\}$ is a set of elements in D such that each a_{i+1} is a proper divisor of a_i, then the set is finite.

Proof: Assume the contrary. Since each a_{i+1} is a proper divisor of a_i, $Da_i \subsetneqq Da_{i+1}$ (Exercise 7, page 184). We obtain a contradiction by proving that for some $n \in N$, $Da_{n+1} \subset Da_n$. Let $\mathscr{I} = \{Da_i \mid a_i \in A\}$ and set

(18) $$I = \bigcup \mathscr{I};$$

by Theorem 24, page 169, I is an ideal in D. Since D is a PID, there is an element $b \in D$ such that

$$I = bD,$$

and moreover $b \in I$. By (18), there is an $n \in N$ such that $b \in Da_n$; hence $I = Db \subset Da_n$. But for each $i \in N$, $Da_i \subset I$, hence, in particular, $Da_{n+1} \subset I$. From the preceding inclusion, it follows that $Da_{n+1} \subset Da_n$, and this is the desired contradiction.

Q.E.D.

THEOREM 34a. a_0 is a product of primes.

Proof: Assume for the moment we have proved: If a_0 is not a product of primes, then a_0 has a proper divisor a_1 which is also not a product of primes. Since a_1 is nonzero and a nonunit, from this assumption it follows that a_1 has a proper divisor a_2 which is again not a product of primes. In turn, a_2 has a proper divisor a_3 which is not a product of primes. Continuing, we obtain a set of elements $\{a_1, a_2, \cdots, a_i, a_{i+1}, \cdots\}$ such that each a_{i+1} is a proper divisor of a_i, $i = 1, 2, \cdots$. By Lemma 1, no such set exists. Hence, a_0 is a product of primes.

Now to prove the assumption valid. If a_0 is not a product of primes, then $a_0 = bc$, where b and c are proper divisors of a_0. If both b and c were products of primes, a_0 would be a product of primes, a contradiction. Therefore, at least one of b, c is not a product of primes. This completes the proof of the theorem.

Q.E.D.

Our next task is to prove that every prime factorization in D is unique to within order of the factors, and to within units. The first steps are:

LEMMA 2. If p and a are elements in D, where p is a prime and $p \nmid a$, then g.c.d. $(p, a) = e$.

Proof: Exercise.

LEMMA 3. If p, a and b are elements in D such that p is a prime and $p \mid ab$, then $p \mid a$ or $p \mid b$.

Proof: It suffices to show that if $p \nmid a$, then $p \mid b$. By Lemma 2 and Exercise 1, page 185, there exist elements s and t in D such that

$$e = sp + ta.$$

Then

$$b = eb = spb + tab;$$

since $p \mid ab$, it follows that $p \mid b$.

Q.E.D.

Now the uniqueness theorem.

THEOREM 34b. If a is an element in D such that

$$a = \epsilon p_1^{m_1} p_2^{m_2} \cdots p_r^{m_r} = \eta q_1^{n_1} q_2^{n_2} \cdots q_t^{n_t},$$

where
(a) the p's and q's are primes,
(b) for $i \neq j$, p_i and p_j are nonassociates, and likewise q_i and q_j are non-associates,
(c) ϵ and η are units in D, and
(d) the m's and n's are positive integers,
then
(i) $m_1 + m_2 + \cdots + m_r = n_1 + n_2 + \cdots + n_t$;
(ii) for each p_i, there is a q_j such that p_i and q_j are associates and $m_i = n_j$;
(iii) for each q_j, there is a p_i such that q_j and p_i are associates and $m_i = n_j$.

Remark: The assumption that the p's and q's are all primes is essential. Indeed, if primeness of the factors is not required, then in the domain of integers

$$12 = 4 \cdot 3 = 6 \cdot 2 = 3^2 \cdot 2^2 = (-3)(-4) = \text{etc.}$$

Before turning to the proof of Theorem 34b, let us consider an informal argument which exposes the ideas underlying the proof. Suppose

$$a = p_1 p_2 \cdots p_m = q_1 q_2 \cdots q_n,$$

where for simplicity it is assumed that $p_1, p_2, \cdots, p_m$ are distinct primes, and $q_1, q_2, \cdots, q_n$ are likewise distinct primes, and where $m \geq n$. Since $p_1 | p_1 p_2 \cdots p_m$, we know that $p_1 | q_1 q_2 \cdots q_n$. By Lemma 3 and induction,

$$p_1 | \text{ one of the } q\text{'s, say, } p_1 | q_1.$$

But since q_1 is prime,

$$q_1 = \epsilon_1 p_1,$$

where ϵ_1 is a unit. Therefore

$$p_1 p_2 \cdots p_m = (\epsilon_1 p_1) q_2 \cdots q_n,$$

and so, by cancellation,

$$p_2 \cdots p_m = \epsilon_1 q_2 \cdots q_n = q_2' q_3 \cdots q_n,$$

where $q_2' = \epsilon_1 q_2$. Similarly for p_2 and, say, q_2', we get

$$p_3 \cdots p_m = q_3' q_4 \cdots q_n.$$

Now if $m > n$, then we should run out of q's before the p's are exhausted, leaving us with

$$p_{n+1} \cdots p_m = \text{unit},$$

a contradiction. Hence $m = n$ and the theorem follows.

The interested reader may now turn to the detailed proof of Theorem 34b.

> *Proof:* If D contains no primes, the theorem is true. Next, assume D contains primes and proceed by induction on $s = m_1 + m_2 + \cdots + m_r$. Let
>
> $K = \{m \mid m \geq 1$, and if $a = \epsilon p_1^{m_1} p_2^{m_2} \cdots p_r^{m_r} = \eta q_1^{n_1} q_2^{n_2} \cdots q_t^{n_t}$ (ϵ and η units in D) where $m = m_1 + m_2 + \cdots + m_r$ and $n = n_1 + n_2 + \cdots + n_t$, then $m = n$; for each p_i there is a q_j such that $p_i = \epsilon_j q_j$, where ϵ_j is a unit in D and $m_i = n_j$; for each q_j there is a p_i and a unit ϵ_j in D such that $p_i = \epsilon_j q_j$ and $m_i = n_j\}$.
>
> First, $1 \in K$. For if $a = \epsilon p_1 = \eta q_1^{n_1} q_2^{n_2} \cdots q_t^{n_t}$ where p_1 and the q's are primes, and if $n = n_1 + n_2 + \cdots + n_t > 1$, then q_1 is a proper divisor of p_1, a contradiction. Therefore $n = 1$ and $a = \epsilon p_1 = \eta q_1$.
>
> Next, suppose $m \in K$ where $m \geq 1$; we prove that $m + 1 \in K$. Let
>
> $$b = \epsilon p_1^{m_1} p_2^{m_2} \cdots p_v^{m_v} = \eta q_1^{n_1} q_2^{n_2} \cdots q_t^{n_t}$$
>
> where the p's and q's are primes and $m_1 + m_2 + \cdots + m_v = m + 1$. Since $p_1 \mid b$, we deduce $p_1 \mid q_1^{n_1} q_2^{n_2} \cdots q_t^{n_t}$. By Lemma 3, $p_i \mid$ some q_j (details ?), say, $p_1 \mid q_1$. Since p_1 and q_1 are prime, $p_1 = \eta_1 q_1$ where η_1 is a unit in D. Setting $\epsilon_1 = \epsilon \eta_1^{m_1}$, ϵ_1 is a unit in D and we have
>
> $$b = \epsilon p_1^{m_1} p_2^{m_2} \cdots p_v^{m_v} = \epsilon_1 q_1^{m_1} p_2^{m_2} \cdots p_v^{m_v} = \eta q_1^{n_1} q_2^{n_2} \cdots q_t^{n_t};$$
>
> by the cancellation law for multiplication in D,
>
> $$c = \epsilon_1 q_1^{m_1-1} p_2^{m_2} \cdots p_v^{m_v} = \eta q_1^{n_1-1} q_2^{n_2} \cdots q_t^{n_t}.$$
>
> Here $(m_1 - 1) + m_2 + \cdots + m_v = m$, so by the induction hypothesis, and because we already have $p_1 = \eta_1 q_1$, we deduce
>
> $$p_1 = \eta_1 q_1 \text{ and } m_1 - 1 = n_1 - 1;$$
>
> for each $i \neq 1$, $p_i =$ some $\epsilon_j q_j$, where ϵ_j is a unit in D and $m_i = n_j$;
>
> for each $j \neq 1$, there is a p_i and a unit ϵ_j in D such that
>
> $$p_i = \epsilon_j q_j \text{ and } m_i = n_j, \text{ and}$$
>
> $(m_1 - 1) + m_2 + \cdots + m_v = m = (n_1 - 1) + n_2 + \cdots + n_t$.
>
> Therefore $m_1 = n_1$ and $m + 1 = n_1 + n_2 + \cdots + n_t$.
>
> Q.E.D.

DEFINITION 23.   An integral domain D is a *unique factorization domain* (UFD) if and only if

(i) for each $a \in D$, $a \neq 0$ and $a \neq$ unit, there exist primes $p_1, p_2, \cdots, p_r$, $r \geq 1$, positive integers $m_1, m_2, \cdots, m_r$, and a unit ϵ in D such that

$$a = \epsilon p_1^{m_1} p_2^{m_2} \cdots p_r^{m_r};$$

and if

(ii) $\epsilon p_1^{m_1} p_2^{m_2} \cdots p_r^{m_r} = \eta q_1^{n_1} q_2^{n_2} \cdots q_t^{n_t}$ where ϵ and η are units in D, the m's and n's are positive integers, the p's and q's are primes, and for $i \neq j$, p_i and p_j are nonassociates and also q_i and q_j are nonassociates, then: $r = s$; for each p_i, there is a q_j such that $p_i = \epsilon_j q_j$, ϵ_j a unit in D; for each q_j there is a p_i and a unit ϵ_j such that $p_i = \epsilon_j q_j$.

Theorems 34a and 34b are now summarized by

THEOREM 35.   Every PID is a UFD.

Two questions: (1) Do there exist integral domains which are not UFD? (2) Do there exist UFD's which are not PID's? The second question will be answered in Chapter 5. To answer the first, we give

EXAMPLE 19.   Let $Z[\sqrt{-5}] = \{a + b\sqrt{-5} \mid a, b \in Z\}$ and define sum and product for $Z[\sqrt{-5}]$ by

$$(a + b\sqrt{-5}) + (c + d\sqrt{-5}) = (a + c) + (b + d)\sqrt{-5}$$

and

$$(a + b\sqrt{-5})(c + d\sqrt{-5}) = (ac - 5bd) + (ad + bc)\sqrt{-5},$$

respectively. $Z[\sqrt{-5}]$ is an integral domain containing Z as a subdomain. We prove that $Z[\sqrt{-5}]$ is not a UFD, hence (Theorem 35) it is not a PID. Define a *norm* N on $Z[\sqrt{-5}]$, $N: Z[\sqrt{-5}] \longrightarrow Z^+$, by

$$N(\alpha) = \alpha\alpha^*, \quad \alpha \in Z[\sqrt{-5}],$$

where, if $\alpha = a + b\sqrt{-5}$, then $\alpha^* = a - b\sqrt{-5}$. Thus $N(\alpha) = a^2 + 5b^2$, and clearly $N(\alpha) \geq 0$; $N(\alpha) = 0$ if and only if $\alpha = 0$. Also, $N(\alpha\beta) = N(\alpha)N(\beta)$.

(i) α is a unit in $Z[\sqrt{-5}]$ if and only if $N(\alpha) = 1$; the only units in $Z[\sqrt{-5}]$ are 1 and -1.

> *Proof:* If α is a unit, then there is an element $\beta \in Z[\sqrt{-5}]$ such that $\alpha\beta = 1$. Then $N(\alpha)N(\beta) = N(\alpha\beta) = N(1) = 1$. Since $N(\alpha) \geq 1$ and $N(\beta) \geq 1$, it follows that $N(\alpha) = N(\beta) = 1$.
>
> Conversely, if $N(\alpha) = 1$ where $\alpha = a + b\sqrt{-5}$, then $a^2 + 5b^2 = 1$, whence $b = 0$ and $\alpha = 1$ or -1. Therefore $\alpha = 1$ or -1 and these are evidently units in $Z[\sqrt{-5}]$.

(ii) For each $\alpha \in Z[\sqrt{-5}]$, $N(\alpha) \neq 3$ and $\neq 7$.

Proof: Exercise.

(iii) $3, 7, 1 + 2\sqrt{-5}$, and $1 - 2\sqrt{-5}$ are primes in $Z[\sqrt{-5}]$.

> *Proof:* We illustrate the proof with 3 and $1 + 2\sqrt{-5}$, and leave the remaining cases as exercises.
>
> Suppose $\alpha\beta = 3$ where α and β are nonunits in $Z[\sqrt{-5}]$. Then $N(\alpha) > 1$, $N(\beta) > 1$ and $N(\alpha)N(\beta) = N(\alpha\beta) = N(3) = 9$. Hence $N(\alpha) = N(\beta) = 3$, contrary to (ii). Therefore 3 is prime.

If α, β are nonunits such that $\alpha\beta = 1 + 2\sqrt{-5}$, then $N(\alpha)N(\beta) = N(\alpha\beta) = N(1 + 2\sqrt{-5}) = 21$, which again contradicts (ii).

(iv) $Z[\sqrt{-5}]$ is not a UFD, hence (Theorem 35) it is not a PID.

Proof: $21 = 7 \cdot 3 = (1 + 2\sqrt{-5})(1 - 2\sqrt{-5})$ where $3, 7, 1 + 2\sqrt{-5}$, $1 - 2\sqrt{-5}$ are primes. Since the only units in $Z[\sqrt{-5}]$ are 1 and -1, the two factorizations do not differ merely by unit factors. Hence the assertion follows.

Exercise: Let p be a prime in a UFD. Prove: If $p \mid ab$, then $p \mid a$ or $p \mid b$.

13. Prime and Maximal Ideals

Every ideal in the domain of integers is principal and is therefore generated by some integer. In particular, if p is a prime number, it is reasonable to call

$$pZ = \{px \mid x \in Z\}$$

a "prime ideal." In this section, we generalize the concept of primeness to ideals in arbitrary rings. The starting point of one generalization is the following observation:

If a product uv, of integers u and v, is in the ideal pZ, where p is a prime, then

$$uv = zp$$

for some $z \in Z$, and therefore

$$p \mid uv.$$

Since p is a prime number, $p \mid u$ or $p \mid v$, which yields, in turn,

$$u \in pZ \text{ or } v \in pZ.$$

Thus the ideal pZ has the property:

(19)   If $uv \in pZ$, then $u \in pZ$ or $v \in pZ$.

Conversely, if $I \neq Z$ is an ideal in Z having property (19), then $I = \{0\}$ or else $I = qZ$, $q > 0$, and in the latter case we assert that q is a prime. Indeed, if q is composite, say, $q = q_1 q_2$ where $1 < q_1 < q$ and $1 < q_2 < q$, then

$$q_1 q_2 = q = q \cdot 1 \in qZ.$$

But since neither q_1 nor q_2 is an integral multiple of q, $q_1 \notin qZ$ and $q_2 \notin qZ$. Thus if q is composite, (19) is violated. Therefore q is prime.

The above discussion leads to one generalization of primeness, namely,

DEFINITION 24. An ideal P in a ring R is a *prime* ideal if and only if for each x and each y in R, if $xy \in P$, then $x \in P$ or $y \in P$.

The introductory remarks suggest

THEOREM 36. In a principal ideal domain D, a proper ideal is prime if and only if it is generated by a prime element or a unit.

Proof: Exercise. Use the analysis of prime ideals in Z as a model.

Exercises

1. Prove that the zero ideal in a ring R is a prime ideal if and only if R has no zero divisors.

2. Prove: In every ring the unit ideal is prime.

3. Verify that the principal ideal (3) is not a prime ideal in $Z[\sqrt{-5}]$ even though 3 is a prime element in $Z[\sqrt{-5}]$. (See Example 19.)

The generalization of primeness of ideals (Definition 24) would be pointless if there did not exist (i) rings which are not integral domains containing proper prime ideals, and (ii) integral domains with prime ideals that are not principal. An instance of (i) is easy. In Z_{12}, $P = \{\bar{0}, \bar{3}, \bar{6}, \bar{9}\}$ is a prime ideal (details?). For an instance of (ii), we return to Example 19, page 190, using all the results already obtained.

Let $P = (3, 1 + 2\sqrt{-5}) = \{3\alpha + (1 + 2\sqrt{-5})\beta \,|\, \alpha, \beta \in Z[\sqrt{-5}]\}$. Evidently P is an ideal in $Z[\sqrt{-5}]$; we shall prove that P is not principal. Note that all the integral multiples of 3 are in P, i.e., $3Z \subset P$. But even more:

(i) The only integers in P are the integral multiples of 3.

Proof: Suppose the integer n is in P; then there exist $a + b\sqrt{-5}$ and $c + d\sqrt{-5}$ such that

$$3(a + b\sqrt{-5}) + (1 + 2\sqrt{-5})(c + d\sqrt{-5}) = n,$$

i.e.,

$$(3a + c - 10d) + (3b + 2c + d)\sqrt{-5} = n.$$

Hence

(20) $3a + c - 10d = n$ and $3b + 2c + d = 0$.

From the second equation of (20), $d = -3b - 2c$, so substituting in the first, one obtains

$$n = 3a + 21c + 30b = 3(a + 7c + 10b).$$

Hence $3 \,|\, n$ and (i) is proved.

(ii) P is not a principal ideal.

Proof: Suppose P is principal with generator $\gamma \in Z[\sqrt{-5}]$. Since 3 and $1 + 2\sqrt{-5}$ are in P, $\gamma \mid 3$ and $\gamma \mid 1 + 2\sqrt{-5}$. But 3 and $1 + 2\sqrt{-5}$ are prime; therefore γ is a unit, i.e., (Example 19) $\gamma = 1$ or $\gamma = -1$. Thus if P is principal, $1 \in P$ and this contradicts (i).

(iii) For each $\alpha \in Z[\sqrt{-5}]$, there is an $n \in Z$ such that $\alpha \equiv n \pmod{P}$.

Proof: The reader should review the definition and properties of $\equiv$ (mod P), pages 175–177.

First note that $2 + \sqrt{-5} \equiv 0 \pmod{P}$; for,

$$2 + \sqrt{-5} = 3(1 + \sqrt{-5}) + (1 + 2\sqrt{-5})(-1) \in P.$$

Then

$$\sqrt{-5} \equiv -2 \pmod{P},$$

whence

$$\alpha = a + b\sqrt{-5} \equiv a - 2b \pmod{P},$$

where $a - 2b \in Z$.

(iv) P is a prime ideal.

Proof: Suppose $\alpha, \beta \in Z[\sqrt{-5}]$ and $\alpha\beta \equiv 0 \pmod{P}$. Let m, n be integers such that

$$m \equiv \alpha \pmod{P} \text{ and } n \equiv \beta \pmod{P}.$$

Then

$$mn \equiv \alpha\beta \pmod{P};$$

but $\alpha\beta \equiv 0 \pmod{P}$ and therefore $mn \equiv 0 \pmod{P}$, i.e., mn is an integer in P. By (i), $3 \mid mn$ whence $3 \mid m$ or $3 \mid n$. In the first case, $m \equiv 0 \pmod{P}$, and since $\alpha \equiv m \pmod{P}$, one deduces $\alpha \equiv 0 \pmod{P}$, i.e., $\alpha \in P$. Similarly, in the second case, $\beta \in P$. By Definition 24, P is a prime ideal.

Now let us consider a second property of ideals generated by prime numbers in the domain of integers. Suppose $P = pZ$, p a positive prime, is such an ideal. If I is an ideal in Z such that

$$P \subsetneqq I,$$

then $I = qZ$ for some integer $q > 0$. But since $p \in P$, $p \in qZ$ and so $p = qz$ for some integer z. Consequently $q \mid p$ and, since p is a prime, $q = 1$ or $q = p$. The latter case is impossible, since otherwise $I = P$. Therefore $q = 1$ and so $I = Z$. Thus the ideal P has the *maximality* property that there is no proper ideal I such that $P \subsetneqq I$. This discussion leads to the following generalization of primeness of integers:

DEFINITION 25. Let M be an ideal in a ring R. M is a *maximal* ideal in R

if and only if
   (i) $M \neq R$, and
   (ii) if I is an ideal such that $M \subset I \subset R$, then $I = M$ or $I = R$.

The remarks preceding Definition 25 suggest

THEOREM 37.   If R is a principal ideal domain, then every prime ideal $\neq$ unit ideal is maximal.

   Proof: Let P be a prime ideal in R, and let I be an ideal such that $P \subset I \subset R$. Since R is a PID with unity element, there exist elements p and q in R such that

$$P = pR \text{ and } I = qR.$$

By Theorem 36, p is a prime element in P. We prove: Either $I = R$ or else $I = P$.
   Since $p \in P$ and $P \subset I, p \in I$, and therefore $p = xq$ for some $x \in R$. Hence $q \mid p$. But since p is a prime, q is a unit or $q = up$ where u is a unit. In the first case $I = qR = R$, and in the second case $I = (pu)R = pR = P$.

                                                                                Q.E.D.

   There are many rings containing prime ideals which are not maximal; obviously such rings cannot be PID's with unity elements. Examples will have to wait until Chapter 5.
   In the study of the modular arithmetics Z_n, we found that Z_n is a field if and only if n is a prime number (Theorem 14). Since Z_n is the residue class ring of Z *modulo* the ideal nZ, it follows that Z_n is a field if and only if nZ is a prime ideal. Now since Z is a PID, every prime ideal is also maximal. Thus Z_n is a field if and only if the ideal nZ is a maximal ideal in Z. We are now led to conjecture: R/I is a field if and only if I is a maximal ideal in R. With but one additional hypothesis, this conjecture is correct. The more immediate guess—R/I is a field if and only if I is a prime ideal—is false. But counterexamples are not available until Chapter 5, where we give a prime ideal which is not maximal. The main theorem of this section is

THEOREM 38.   Let R be a ring with unity element e, and let I be an ideal in R. Then the residue class ring R/I
   (i) is a field if and only if I is a maximal ideal in R;
   (ii) is an integral domain if and only if I is a prime ideal in R.

   Proof: (i) Suppose I is a maximal ideal in R. Since R has a unity element, and since $I \neq R$, R/I has $\bar{e} = e + I$ as unity element (Theorem 29) and $\bar{e} \neq \bar{0}$ (Exercise 4, page 177). Let $\bar{a} \in R/I$ where $\bar{a} = a + I \neq \bar{0}$; our object is to find an element $b \in R$ such that

$$\bar{b} \odot \bar{a} = \bar{e}.$$

Since $\bar{a} = a + I \neq \bar{0} = I$, $a \notin I$. Now let A be the ideal in R generated by the set $I \cup \{a\}$. By Exercise 6, page 174, $I \subset A \subset R$. Since I is a maximal ideal in R, $A = R$ and so $e \in A$. Again, by Exercise 6, page 174, there are elements u and b in I and R, respectively, such that

$$e = u + ba.$$

Then

$$\bar{e} = \overline{u + ba} = \bar{u} \oplus (\bar{b} \odot \bar{a});$$

but since $u \in I$, $\bar{u} = \bar{0}$, and therefore

$$\bar{b} \odot \bar{a} = \bar{e}.$$

Conversely, suppose R/I is a field. To prove that I is a maximal ideal in R, let B be an ideal in R such that $I \subsetneqq B \subset R$. We show that $e \in B$, whence $B = R$ and therefore I is maximal. Since $I \subsetneqq B$, there is an $x \in B$ such that $x \notin I$. Then $\bar{x} \neq \bar{0}$ and so there is a $y \in R$ such that

$$\bar{x} \odot \bar{y} = \bar{e}.$$

Hence $xy - e \in I \subset B$. But $x \in B$ implies $xy \in B$ and therefore $e \in B$. This completes the proof of (i).

(ii) Suppose I is a prime ideal; we show that R/I contains no divisors of zero, and therefore it is an integral domain. Let $\bar{x}$, $\bar{y} \in R/I$, where $\bar{x} = x + I$, $\bar{y} = y + I$, and $\bar{x} \odot \bar{y} = \bar{0}$. Then

$$\bar{x} \odot \bar{y} = (x + I) \odot (y + I) = xy + I = \bar{0} = I,$$

whence $xy \in I$. But since I is a prime ideal, $xy \in I$ implies $x \in I$ or $y \in I$. In the former case, $\bar{x} = \bar{0}$, and in the latter case, $\bar{y} = \bar{0}$. Thus R/I contains no divisors of zero, hence is an integral domain.

Conversely, let R/I be an integral domain. If I is not a prime ideal, there exist x and y in R such that $xy \in I$, but $x \notin I$ and $y \notin I$. But $xy \in I$ implies $\overline{xy} = \bar{x} \odot \bar{y} = \bar{0}$; on the other hand, $x \notin I$ and $y \notin I$ yield $\bar{x} \neq \bar{0}$ and $\bar{y} \neq \bar{0}$, respectively. Hence $\bar{x}$ and $\bar{y}$ are divisors of zero in R/I, a contradiction. Therefore I is a prime ideal.

Q.E.D.

COROLLARY 1. A maximal ideal in a ring with unity element is a prime ideal.

Proof: Exercise.

COROLLARY 2. Let R be a commutative ring with unity element, and let η be a homomorphism of R.

(i) $\eta[R]$ is a field if and only if $\ker(\eta)$ is a maximal ideal in R;

(ii) $\eta[R]$ is an integral domain if and only if $\ker(\eta)$ is a prime ideal in R.

Proof: Exercise.

Q.E.D.

Exercises

1. Without using the result of Theorem 38, prove that in a ring with unity element every maximal ideal is prime.

2. As usual, let G be the domain of Gaussian integers. Use the results of Example 17 and Theorems 37 and 38 to prove that $G/3G$ is a field. (*Hint:* First prove that 3 is a prime in G.)

We conclude this section by presenting a table which compares elementary number-theoretic notions of divisibility with the ideal-theoretic view of these same concepts.

TABLE 15

Divisibility	Ideal Theory
$a \mid b$	$(a) \supset (b)$
u is a unit	$(u) = D$
a and b are associates	$(a) = (b)$
$d = $ g.c.d. (a, b)	$(d) = (a, b)$, i.e., the principal ideal generated by d is the ideal generated by $\{a, b\}$
prime number	prime ideal; and if D is a PID, maximal ideal
$p \mid ab$ implies $p \mid a$ or $p \mid b$	$ab \in P$, a prime ideal, implies $a \in P$ or $b \in P$

The reader should try to extend this table by listing other correspondences of the same kind.

14. *The Quotient Field of an Integral Domain*

In the systematic study of elementary arithmetic, one proceeds from the integers to the construction of the field of rational numbers (see Hamilton and Landin, *Set Theory and the Structure of Arithmetic*, Chapter 3, pages 157–161). An examination of this process shows that the construction depends only upon the fact that Z is an integral domain with unity element and uses no other properties of Z. Thus it must be true that given an integral domain D with unity element, one can construct, in a similar fashion, a field Q of quotients of elements in D. In such a field, one can carry out unrestricted division by nonzero elements. The first step in the proof of our existence theorem is a technical lemma[3] which is essential to the proof of the main result, and will also be used for other purposes. (See Theorem 3, page 208.)

---

[3] It is suggested that the proof of the lemma be omitted in a first reading of this text.

LEMMA. Let $\sigma : R \longrightarrow R'$ be an isomorphism between rings R and R', and let R be a subring of a ring S. Then there exist an extension ring S' of R' and an isomorphism $\tau : S \longrightarrow S'$ such that $\tau \,|\, R = \sigma$.

Proof: The basic idea in the proof of the lemma is to obtain S' from S by replacing the subset R by R'. The procedure requires case distinctions.

CASE 1. $(S - R) \cap R' = \emptyset$. Set $S' = (S - R) \cup R'$ and define $\tau : S \longrightarrow S'$ by

$$(21) \qquad \tau(x) = \begin{cases} x, & \text{if } x \in S - R, \\ \sigma(x), & \text{if } x \in R. \end{cases}$$

Clearly τ is a bijection such that $\tau \,|\, R = \sigma$. Next, assuming that sum and product in S (and in R) are denoted by "$+$" and "$\cdot$" (or, juxtaposition), respectively, define $\oplus$ and $\odot$ on S', thus:

If $a', b' \in S'$, then there exist unique $a, b \in S$ such that $\tau(a) = a'$ and $\tau(b) = b'$. Set

$$a' \oplus b' = \tau(a + b) \quad \text{and} \quad a' \odot b' = \tau(ab).$$

By their definitions, both $\oplus$ and $\odot$ are mappings $S' \times S' \longrightarrow S'$, hence they are binary operations on S'.

With respect to $\oplus$, S' is an Abelian group. Indeed, if $a', b', c' \in S'$ and if a, b, c are the unique elements in S such that $\tau(a) = a'$, $\tau(b) = b'$ and $\tau(c) = c'$, then

$$\begin{aligned} a' \oplus (b' \oplus c') &= \tau(a) \oplus \tau(b + c) \\ &= \tau(a + (b + c)) \\ &= \tau((a + b) + c) \\ &= \tau(a + b) \oplus \tau(c) \\ &= (a' \oplus b') \oplus c'. \end{aligned}$$

By similar techniques the reader may verify that $0' = \tau(0)$ is the zero element of S', that $\tau(-a)$ is the additive inverse of $a' = \tau(a)$, and that $\oplus$ is commutative. Therefore S' is an Abelian group with respect to $\oplus$.

The reader may also verify that $\odot$ is associative. Finally, $\odot$ is distributive over $\oplus$. For, with the notations introduced above,

$$\begin{aligned} a' \odot (b' \oplus c') &= \tau(a) \odot \tau(b + c) \\ &= \tau(a(b + c)) \\ &= \tau(ab + ac) \\ &= \tau(ab) \oplus \tau(ac) \\ &= \tau(a) \odot \tau(b) \oplus \tau(a) \odot \tau(c) \\ &= a' \odot b' \oplus a' \odot c'. \end{aligned}$$

Therefore S' is a ring and by equation (21) the lemma is proved in Case 1.

CASE 2. $(S - R) \cap R' \neq \emptyset$. In this case the method of setting $S' = (S - R) \cup R'$ and defining $\tau : S \longrightarrow S'$ by equations (21) runs into difficulties (what are they?). We therefore seek to apply the method of Case 1 by first replacing R and S by rings R^* and S^*, respectively, such that $(S^* - R^*) \cap R' = \emptyset$.

To effect the replacement, first note that among the elements in R' there may be some which are ordered pairs, others which are not. Now *let u be an element which is not a first component of an ordered pair in R',* and set

$$S^* = \{u\} \times S;$$

then $R^* = \{u\} \times R$ is a subset of S^*. Using the same symbols, "+" and "·" for sum and product in S^* as in S, we define

$$(u, x) + (u, y) = (u, x + y),$$

and

$$(u, x)(u, y) = (u, xy),$$

where $x, y \in S$. If we further define $\eta : S \longrightarrow S^*$ by

$$\eta(x) = (u, x), \ x \in S,$$

then obviously η is an isomorphism such that $\eta[S] = S^*$ and $\eta[R] = R^*$. Therefore S^* is a ring having R^* as a subring. Moreover, the choice of the element u shows that

$$(S^* - R^*) \cap R' = \emptyset.$$

If we let $\mu = \eta^{-1} \,|\, R^*$, the situation may be pictured in Figure 17.

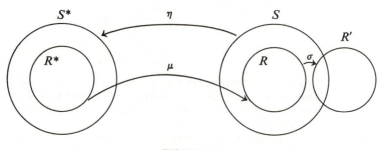

FIGURE 17

By the definition of μ, $\sigma \circ \mu : R^* \longrightarrow R'$ is an isomorphism. By Case 1 there is an extension ring S' of R' and an isomorphism $\rho : S^* \longrightarrow S'$ such that $\rho \,|\, R^* = \sigma \circ \mu$, i.e., $\rho[R^*] = R'$. Then $\rho \circ \eta : S \longrightarrow S'$ is an isomorphism and

$$\begin{aligned}(\rho \circ \eta)[R] &= \rho[\eta[R]] \\ &= \rho[R^*] \\ &= R' \\ &= \sigma[R],\end{aligned}$$

so that $\rho \circ \eta \,|\, R = \sigma$. Thus, S' and $\rho \circ \eta$ are the desired extension ring and isomorphism, respectively. This completes the proof in Case 2.

Q.E.D.

We remark, in passing, that if S has a unity element, so does S', and if S is commutative, so is S'.

Exercise: Prove that Z_3 has a nine-element extension field B which is isomorphic to $G/3G$ (Example 16, page 177). Show that each element of B has the form $a + b\alpha$ where $a, b \in Z_3$, $\alpha \in B$, and $\alpha^2 = \bar{2}$.

We now use the Lemma to establish the main result. Let us recall that the integers Z comprise a subdomain of the field of rational numbers Q, and that for each $r \in Q$, there exist $a, b \in Z$, $b \neq 0$, such that

$$a = br.$$

Whenever a field exhibits this property relative to the subdomain, we speak of the field as a "quotient field" of the given subdomain. More precisely,

DEFINITION 26.   Let D be an integral domain with unity element e. An extension field F of D is a *quotient field* of D if and only if for each $q \in F$ there exist $a, b \in D$, $b \neq 0$, such that

$$a = bq.$$

THEOREM 39.   Every integral domain D with unity element e has a quotient field.

Proof: We shall construct a field F containing a subdomain E and a mapping $\sigma : E \longrightarrow D$ such that

   (i) σ is an isomorphism, and
   (ii) F is a quotient field of E.

By the lemma it will follow that there is an extension field F' of D and an isomorphism $\tau : F \longrightarrow F'$ such that $\tau \,|\, E = \sigma$. By Exercise 1, page 166, F' is a field. We shall then verify that F' is a quotient field of D.

   Let $D_0 = D - \{0\}$; define a relation $\sim$ on $D \times D_0$ by

$$(a, b) \sim (c, d) \text{ if and only if } ad = bc,$$

where (a, b), (c, d) are elements in $D \times D_0$. Evidently $\sim$ is reflexive and symmetric. If $(a, b) \sim (c, d)$ and $(c, d) \sim (e, f)$, then

$$ad = bc \quad \text{and} \quad cf = de,$$

whence

$$adf = bcf \quad \text{and} \quad bcf = bde,$$

therefore

$$adf = bde;$$

since $d \neq 0$ and D is an integral domain, $af = be$, hence (a, b)

$\sim (e, f)$. Thus $\sim$ is transitive and so it is an equivalence relation on $D \times D_0$. Therefore $\sim$ determines a partition F of $D \times D_0$. The reader will recall that if $A \in F$ and if $(a, b) \in A$, then

(22) $A = \{(x, y) \mid (x, y) \in D \times D_0 \text{ and } (x, y) \sim (a, b)\};$

henceforth we write

$$A = \frac{a}{b}.$$

By (22)

$$\frac{a}{b} = \frac{c}{d} \text{ if and only if } (a, b) \sim (c, d),$$

hence if and only if $ad = bc$.

Our next step is to prove that with appropriately defined sum and product, F is a field. Observe, first, that if a/b, $c/d \in F$, then $ad + bd$ and ac are elements in D and $bd \in D_0$. Consequently $(ad + bc)/bd$ and ac/bd are elements in F. Further, if

$$\frac{a}{b} = \frac{e}{f} \text{ and } \frac{c}{d} = \frac{g}{h},$$

then a computation shows that

$$\frac{ad + bc}{bd} = \frac{eh + fg}{fh},$$

and

$$\frac{ac}{bd} = \frac{eg}{fh}.$$

Now define $\oplus$ and $\odot$ on F by

$$\frac{a}{b} \oplus \frac{c}{d} = \frac{ad + bc}{bd},$$

and

$$\frac{a}{b} \odot \frac{c}{d} = \frac{ac}{bd},$$

respectively. The foregoing discussion shows at once that $\oplus$ and $\odot$ are binary operations on F.

To prove that F is a field requires several verifications, of which all are left to the reader except the proof that each nonzero element in F has a reciprocal in F. In order to prove the last assertion, note that if $x/y \in F$, then (a) x/y is the unity element of F if and only if $x = y$, and (b) x/y is the zero element of F if and only if $x = 0$.

Suppose, now, x/y is not the zero element of F; then $x \neq 0$ and so $y/x \in F$. But

$$\frac{y}{x} \odot \frac{x}{y} = \frac{yx}{xy} = \frac{xy}{xy};$$

since $xy \neq 0$, xy/xy is the unity element of F, and therefore y/x is the reciprocal of x/y.

Now let $E = \{a/e \mid a \in D$ and $e =$ unity element of $D\}$. If a/e, b/e are elements in E, then

$$\frac{a}{e} - \frac{b}{e} = \frac{a - b}{e},$$

and

$$\frac{a}{e} \odot \frac{b}{e} = \frac{ab}{e},$$

so that $a/e - b/e$ and $a/e \cdot b/e$ are in E. Hence E is a subring of F. Define $\sigma : E \longrightarrow D$ by

$$\sigma\left(\frac{a}{e}\right) = a, \quad \frac{a}{e} \in E.$$

Obviously, $a/e = b/e$ if and only if $a = b$ and therefore σ is one-one. The definition of σ shows, in fact, that it is a surjection. Moreover,

$$\sigma\left(\frac{a}{e} \oplus \frac{b}{e}\right) = \sigma\left(\frac{a + b}{e}\right)$$
$$= a + b$$
$$= \sigma\left(\frac{a}{e}\right) + \sigma\left(\frac{b}{e}\right),$$

and similarly,

$$\sigma\left(\frac{a}{e} \odot \frac{b}{e}\right) = \sigma\left(\frac{a}{e}\right)\sigma\left(\frac{b}{e}\right).$$

Hence σ is an isomorphism and E must be an integral domain.

By the lemma there exists an extension field F' of D and an isomorphism $\tau : F \longrightarrow F'$ such that $\tau \mid E = \sigma$. Our theorem is completed by showing that F' is a quotient field of D. To this end, let $v \in F'$. Then there is an $x/y \in F$ such that $\tau(x/y) = v$. Also x/e and y/e are elements in E (with $y \neq 0$) such that

$$\frac{x}{e} = \frac{y}{e} \odot \frac{x}{y}.$$

Applying τ, we have

$$\tau\left(\frac{x}{e}\right) = \tau\left(\frac{y}{e} \odot \frac{x}{y}\right) = \tau\left(\frac{y}{e}\right)\tau\left(\frac{x}{y}\right),$$

or

$$x = yv.$$

Since $x, y \in D$ and $y \neq 0$, F' satisfies the conditions of Definition 26. This completes the proof of the theorem.

Q.E.D.

Remark: If F is a quotient field of an integral domain D, then each element in F is a product of an element in D by the reciprocal of an element in D. Indeed, if $x \in F$, then there exist $a, b \in D$, $b \neq 0$, such that

$$a = bx.$$

But since $D \subset F$, $b \in F$; and since F is a field, $b^{-1} \in F$. Hence

$$x = ab^{-1}.$$

One question regarding quotient fields remains to be settled. To what extent is a quotient field of an integral domain determined? The exercises below yield the result that F is determined to within isomorphism. To speak of *the* quotient field of D merely implies this degree of uniqueness.

Exercises

1. Let F be a quotient field of an integral domain D, and let $\iota : D \longrightarrow F$ be the inclusion monomorphism. Further, let K be a field and let $\eta : D \longrightarrow K$ be a monomorphism. Prove that there exists a monomorphism $\sigma : F \longrightarrow K$ such that $\sigma \circ \iota = \eta$. (*Hint:* For each $ab^{-1} \in Q$ set $\sigma(ab^{-1}) = \eta(a)(\eta(b))^{-1}$, and prove that (a) if $x \in D$, then $(\sigma \circ \iota)(x) = \eta(x)$; (b) σ is a monomorphism.

2. Use the result of Exercise 1 to prove that all quotient fields of D are isomorphic.

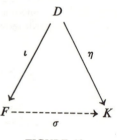

FIGURE 18

5

Polynomial Rings

1. *Introduction; The Concept of Polynomial Ring*

The polynomials studied in abstract algebra constitute a generalization of the polynomials of elementary—high school and college—algebra. We begin at once with an investigation of polynomials, in the spirit of modern algebra. Later (Section 7) we shall prove explicitly that the modern algebraist's version of polynomial is indeed a generalization of the elementary concept.

Let B be a (commutative) ring with unity element, 1, and let A be a *unitary* subring of B, i.e., A is a subring of B having 1 as its unity element. Further, let $t \in B$.

DEFINITION 1. The subring of B generated by the set $A \cup \{t\}$ is the *ring of polynomials in t over A* and it is denoted by "$A[t]$." Each element in $A[t]$ is a *polynomial in t with coefficients in A*.

From the formal definition and from Exercise 5, page 163, it follows that an element $p \in B$ is a polynomial in t with coefficients in A if and only if

$$(1) \qquad p = x_1 x_2 \cdots x_{m_1} + y_1 y_2 \cdots y_{m_2} + \cdots + z_1 z_2 \cdots z_{m_n},$$

where the x's, the y's, $\cdots$, and the z's are elements in $A \cup \{t\}$. For instance, the first term of (1) might be

$$(2) \qquad a_1 \cdots a_{k_1-1} t a_{k_1+1} \cdots a_{k_2-1} t a_{k_2+1} \cdots a_r,$$

where the factor t occurs, say, s times, $s \geq 0$, and the a's are elements in A. In view of the fact that B is commutative and the a's are all elements in A, (2) may be simplified to

$$(3) \qquad at^s,$$

where $a \in A$. Since each term of (1) can be similarly simplified,

(4) $$p = a_0 + a_1 t + \cdots + a_n t^n,$$

where $n \geq 0$ and $a_i \in A$, $0 \leq i \leq n$. A standard abbreviation for the right-hand side of (4) is the expression

$$\sum_{i=0}^{n} a_i t^i,$$

so that we shall frequently use

(4') $$p = \sum_{i=0}^{n} a_i t^i.$$

EXAMPLE 1.   Let $A = Q$ be the field of rational numbers, let $B = R$ be the field of real numbers, and let $t = \pi$ be the famous real number which is the ratio of the circumference to the diameter of a circle. Then $Q[\pi]$ is the subring of R consisting of all polynomials in π with rational coefficients; thus $Q[\pi]$ is the subring of R consisting of all sums

$$a_0 + a_1 \pi + \cdots + a_n \pi^n,$$

where $n \in N$ and $a_i \in Q$, $0 \leq i \leq n$.

EXAMPLE 2.   With A and B as in Example 1, take $t = \sqrt{2}$. Then $Q[\sqrt{2}]$ is the ring of all polynomials

$$a_0 + a_1 \sqrt{2} + \cdots + a_n (\sqrt{2})^n,$$

where $n \in N$ and the a_i are rational numbers. (See Exercise 1 on page 206.)

EXAMPLE 3.   Let $A = Z_3$ and let B be the nine-element field described in the Exercise, page 183. It is then clear that

$$A[\gamma] = B.$$

Exercise: Prove: (i) $A[t]$ is a unitary subring of B and (ii) A is a subring of $A[t]$.

2. *Indeterminates*

Let us observe a crucial difference among Examples 1-3. The real number π (Example 1) is a *transcendental number* and this means: If $a_0, a_1, \ldots, a_n$ are rational numbers, $n \in N$, and if

$$a_0 + a_1 \pi + \cdots + a_n \pi^n = 0,$$

then

$$a_0 = a_1 = \cdots = a_n = 0.$$

The proof of the transcendence of π lies beyond the scope of our study[1] and we shall take this result for granted. By contrast, in Example 2, one finds that

$$6 - 1(\sqrt{2})^2 - 1(\sqrt{2})^4 = 0,$$

where $n = 4$, $a_0 = 6$, $a_1 = 0$, $a_2 = -1$, $a_3 = 0$, $a_4 = -1$. And in Example 3

$$\bar{1} + \gamma^2 = 0,$$

where $n = 2$, $a_0 = \bar{1}$, $a_1 = \bar{0}$, $a_2 = \bar{1}$.

Our examples show that the polynomial rings $A[t]$ fall into two mutually exclusive classes. The first class consists of those polynomial rings in which the element t has the property:

(*) $$\sum_{i=0}^{n} a_i t^i = 0, n \in N, \text{ implies each } a_i = 0, 0 \leq i \leq n.$$

The second class consists of polynomial rings having elements $\sum_{i=0}^{n} a_i t^i$, $n \in N$, such that

$$\sum_{i=0}^{n} a_i t^i = 0, \text{ but some } a_j \neq 0, 0 \leq j \leq n.$$

Since elements t with property (*) will occupy a central role in our theory, it is convenient to introduce a term for them.

DEFINITION 2.   Let $A[t]$ be a polynomial ring in t over A. The element t is an *indeterminate over A* if and only if

$$\sum_{i=0}^{n} a_i t^i = 0, n \in N, \text{ implies each } a_i = 0, 1 \leq i \leq n.$$

Thus π is an indeterminate over the rationals, whereas $\sqrt{2}$ is not. Also, the element γ (Example 3) is not an indeterminate over the modular arithmetic Z_3.

THEOREM 1.   Let t be an indeterminate over the ring A, and let $\sum_{i=0}^{r} b_i t^i$, $\sum_{j=0}^{s} c_j t^j$ be elements in $A[t]$ where b_r and c_s are nonzero. Then

$$\sum_{i=0}^{r} b_i t^i = \sum_{j=0}^{s} c_j t^j$$

---

[1] The interested reader may consult: Ivan Niven, "Irrational Numbers," *The Carus Mathematical Monographs*, Number 11, The Mathematical Association of America, 1956.

if and only if

$$r = s \text{ and } b_i = c_i, \ 0 \leq i \leq r.$$

Proof: Exercise.

Theorem 1 provides a second characterization of the indeterminate, one which is very close to the original definition. In Section 8 we shall examine additional characterizations that yield deeper insights into the nature of indeterminates.

Exercises

1. Prove that $Q[\sqrt{2}] = \{a + b\sqrt{2} \mid a, b \in Q\}$. Verify that $1/\sqrt{2}$ and $1/(a + b\sqrt{2})$ (where $a + b\sqrt{2} \neq 0$) are elements in $Q[\sqrt{2}]$. What is meant by rationalizing the denominator of $(c + d\sqrt{2})/(a + b\sqrt{2})$?

2. Is $1/\pi \in Q[\pi]$? Why?

3. Rationalize the denominator of $1/(1 + 3\sqrt{5} + (3\sqrt{5})^2)$.

4. Suppose A is a unitary subring of a ring B with unity element, and let $t \in B$ be an indeterminate over A. Does it follow that t is an indeterminate over B? Why? If C is a unitary subring of A, is t an indeterminate over C?

5. Let t be an indeterminate over the ring A. Prove that for each $a \neq 0$ and each b, $a, b \in A$, $at + b$ is an indeterminate over A.

6. Generalize the result of Exercise 5 by proving: If t is an indeterminate over A, then $\sum_{i=0}^{n} a_i t^i$ is also indeterminate over A, where $a_i \in A$, $0 \leq i \leq n$, $n > 0$, and $a_n \neq 0$.

Definitions 1 and 2 can be extended easily to the case that A and B are noncommutative rings; one merely requires, additionally, that t commute with each element in A. Such extensions of our definitions are important, for example, in the theory of matrix polynomials. If A and B are not assumed commutative, Definition 1 is replaced by

DEFINITION 1′. Let A be a unitary subring of a ring B, and let t be an element in B such that for each $a \in A$, $at = ta$. The subring of B generated by the set $A \cup \{t\}$ is the *ring of polynomials in t over A* and it is denoted by "$A[t]$."

Because t commutes with each $a \in A$, it is easy to see that p is in $A[t]$ if and only if there is an $n \in N$ and $a_0, a_1, \ldots, a_n$ in A such that

$$p = \sum_{i=0}^{n} a_i t^i.$$

As for Definition 2, it may now be accepted precisely as stated, the only change being that A and B are not assumed to be commutative.

For our purposes, A and B are always taken to be commutative rings.
Nevertheless, the reader will note that Theorem 1, as well as Theorems 2 and
3 below, hold for the noncommutative case essentially without modification.

If t is an indeterminate over a ring A, then Exercise 6, page 206, shows
that there are infinitely many indeterminates over A. Indeed, in the case that
A is the field Q, every transcendental number is an indeterminate over Q
and so there exist uncountably many indeterminates over Q[2]. Nevertheless
there is a basic kind of sameness about all indeterminates over a ring A,
as the next theorem shows.

THEOREM 2.   If t and z are indeterminates over a ring A, then the polynomial
rings $A[t]$ and $A[z]$ are isomorphic.

Proof: Define a mapping $\sigma : A[t] \longrightarrow A[z]$ by setting

$$\sigma\left(\sum_{i=0}^{n} a_i t^i\right) = \sum_{i=0}^{n} a_i z^i.$$

Clearly, σ is a one-one correspondence and since σ preserves sums
and products (details?), it is an isomorphism.

Q.E.D.

Exercises

1. Let x and y be indeterminates over the rings A, B (with unity elements),
respectively, and let f be an isomorphism between A and B. Prove that $\sigma : A[x]$
$\longrightarrow B[y]$ defined by

$$\sigma\left(\sum_{i=0}^{n} a_i x^i\right) = \sum_{i=0}^{n} f(a_i) y^i$$

is an isomorphism such that $\sigma \,|\, A = f$.

2. Let $A[x]$ and $B[y]$ be polynomial rings, and let $\sigma : A[x] \longrightarrow B[y]$ be an
isomorphism such that $\sigma \,|\, A$ is an isomorphism between A and B and $\sigma(x) = y$.
Prove that x is an indeterminate over A if and only if y is an indeterminate over B.

3.   *Existence of Indeterminates*

In Section 2 we found that there exist indeterminates over Q; in fact,
every transcendental number is an indeterminate over Q. The purpose of this

---

[2] See Corollary on page 10, E. Kamke, *Theory of Sets*, translated from the second German
edition by F. Bagemihl (New York, Dover Publications, Inc., 1950).

section is to establish the corresponding result for all commutative rings with unity element.

THEOREM 3. For each (commutative) ring A with unity element, 1, there exists an indeterminate t over A.[3]

Proof: The first and by far the longest step in this proof is the construction of a ring A' isomorphic to A, and an indeterminate x over A'. An application of the lemma, page 197, and of Exercise 2, above, then completes the job.

Recall that a *sequence of elements in A* is a mapping $\mu : N \longrightarrow A$; the elements $\mu(k) = a_k \in A$ where $k \in N$ are the *terms* of the sequence, and the sequence μ will be denoted by

$$(a_0, a_1, a_2, \ldots, a_n, a_{n+1}, \ldots).$$

Now let B' be the set of all sequences such that *at most a finite number of terms are nonzero*. For example,

$$(1, 1, \ldots, 1, 1, \ldots)$$

is not an element in B'. On the other hand, the sequences

$$(0, 0, \ldots, 0, 0, \ldots),$$
$$(1, 1, 0, 0, \ldots, 0, 0, \ldots),$$

and

$$(a_0, a_1, a_2, 0, \ldots, 0, 0, \ldots),$$

where $a_0, a_1, a_2 \in A$, are elements in B'.

Next define sum $\oplus$ of elements $\mu, \gamma \subset B'$ by

$$(\mu \oplus \gamma)(n) = \mu(n) + \gamma(n), \quad n \in N.$$

Thus, if

(5) $$\mu = (a_0, a_1, \ldots, a_j, 0, \ldots),$$

and

(6) $$\gamma = (b_0, b_1, \ldots, b_k, 0, \ldots),$$

where, say, $j \geq k$, then

(7) $$\mu \oplus \gamma = (a_0 + b_0, a_1 + b_1, \ldots, a_k + b_k, a_{k+1}, \ldots, a_j, 0, \ldots).$$

Clearly $\oplus$ is a binary operation on B'. The reader may verify easily that B' is an Abelian group with respect to $\oplus$, and its neutral element is $(0, 0, \ldots)$. The inverse of μ is

$$-\mu = (-a_0, -a_1, \ldots, -a_j, 0, \ldots).$$

The definition of multiplication $\odot$ is a little more complicated. If $\mu, \gamma \in B'$, we define

$$(\mu \odot \gamma)(n) = \sum_{k=0}^{n} \mu(k)\gamma(n - k), \quad n \in N.$$

---

[3] The proof of Theorem 3 may be skipped in a first reading of this chapter.

For instance, if μ and γ are the sequences (5) and (6), respectively, then

$$(\mu \odot \gamma)(0) = a_0 b_0;$$
$$(\mu \odot \gamma)(1) = a_0 b_1 + a_1 b_0;$$
$$(\mu \odot \gamma)(2) = a_0 b_2 + a_1 b_1 + a_2 b_0, \text{ etc.}$$

Hence $\mu \odot \gamma$ is the sequence

$$(8) \qquad (a_0 b_0, a_0 b_1 + a_1 b_0, a_0 b_2 + a_1 b_1 + a_2 b_0, \ldots, a_0 b_j + a_1 b_{j-1} \\ + \cdots + a_j b_0, 0, \ldots).$$

A comparison of (8) with the way in which the polynomial functions of elementary algebra multiply will reveal the source of inspiration for the definition of $\odot$. Since $a_i = 0$ for $i > j$ and $b_m = 0$ for $m > k$, it follows that if $r > j + k$, then

$$b_r = b_{r-1} = \cdots = b_{r-j} = 0 \text{ and } a_{j+1} = a_{j+2} = \cdots = a_r = 0.$$

Hence if $r > j + k$, then

$$(\mu \odot \gamma)(r) = a_0 b_r + a_1 b_{r-1} + \cdots + a_j b_{r-j} + a_{j+1} b_{r-j-1} + \\ \cdots + a_r b_0 = 0.$$

Therefore $\mu \odot \gamma \in B'$, and so $\odot$ is a binary operation on B'.

Evidently $\odot$ is commutative. To prove that $\odot$ is also associative, let μ, γ, $\eta \in B'$ where μ and γ are given by (5) and (6), respectively, and $\eta = (c_0, c_1, \ldots, c_s, 0, \ldots)$. It suffices to show that for each $n \in N$,

$$(9) \qquad \sum_{k=0}^{n}\left(\sum_{j=0}^{k} a_k b_{k-j}\right) c_{n-k} = \sum_{k=0}^{n} a_{n-k}\left(\sum_{j=0}^{k} b_j c_{k-j}\right).$$

The left-hand side of (9) is

$$(a_0 b_0)c_n + (a_0 b_1 + a_1 b_0)c_{n-1} + \cdots + (a_0 b_{n-1} + a_1 b_{n-2} \\ + \cdots + a_{n-1} b_0)c_1 + (a_0 b_n + a_1 b_{n-1} + \cdots + a_n b_0)c_0 \\ = \sum_{i+j+k=n} a_i b_j c_k.$$

Similarly, one verifies that the right-hand side of (9) is $\sum_{i+j+k=n} a_i b_j c_k$ and therefore equation (9) follows.

We leave it to the reader to verify that $\odot$ is distributive over $\oplus$ and that the sequence $(1, 0, \cdots)$ is the unity element of B'. To sum up:

B' is a commutative ring with unity element.

We claim that B' contains a subring A' isomorphic with A. Indeed, let A' be the set of all sequences

$$a' = (a, 0, \ldots), \quad a \in A.$$

The mapping $f: A' \longrightarrow A$ defined by

(10) $$f(a') = a, \quad a' \in A',$$

is evidently an isomorphism (details?).

Next, B' is a polynomial ring in an indeterminate x over A'. Let

$$x = (0, 1, 0, \ldots);$$

then

$$x^0 = (1, 0, \ldots),$$
$$x^1 = x = (0, 1, 0, \ldots),$$
$$x^2 = (0, 0, 1, 0, \ldots),$$
$$x^3 = (0, 0, 0, 1, 0, \ldots), \text{ etc.}$$

Thus for each $n \in N$, x^n is the sequence

$$x^n(n) = 1 \text{ and } x^n(k) = 0 \text{ if } k \neq n.$$

Moreover,

$$(0, 0, \ldots, 0, a_k, 0, \ldots) = (a_k, 0, \ldots) \odot (0, 0, \ldots, 0, \overset{k\text{th place}}{1}, 0, \ldots)$$
$$= a_k' \odot x^k,$$

so that

$$(a_0, a_1, \ldots, a_m, 0, \ldots) = a_0' \oplus a_1' \odot x \oplus a_2' \odot x^2 \oplus \cdots \oplus a_m' \odot x^m.$$

This shows that B' is the polynomial ring $A'[x]$.

If we replace "$\oplus$" and "$\odot$" by "$+$" and "$\cdot$" (or, juxtaposition), respectively, then

$$(a_0, a_1, \ldots, a_m, 0, \ldots) = a_0' + a_1'x + \cdots + a_m'x^m = \sum_{i=0}^{m} a_i'x^i.$$

To prove that x is an indeterminate over A', suppose

$$\mu = \sum_{i=0}^{m} a_i'x^i = 0.$$

Since $\sum_{i=0}^{m} a_i'x^i = (a_0, a_1, \ldots, a_m, 0, \ldots)$,

$$\mu = (a_0, a_1, \ldots, a_m, 0, \ldots)$$

is the zero sequence. This means that each $a_i = 0$, $0 \leq i \leq m$, and therefore

$$a_i' = (0, 0, \ldots) = 0', \quad 0 \leq i \leq m.$$

Hence x is an indeterminate over A'.

We are now in the situation depicted by Figure 19, where f is the isomorphism defined by (10). By the lemma of Chapter 4, page 197, there exists an extension ring B of A and an isomorph-

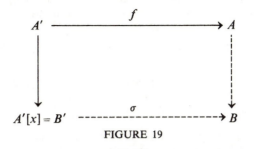

FIGURE 19

ism $\sigma : B' \longrightarrow B$ such that $\sigma \,|\, A' = f$. Letting $t = \sigma(x)$ we check easily that $B = A[t]$. By Exercise 2, page 207, t is an indeterminate over A.

Q.E.D.

Exercises

1. Prove: if A is an integral domain, and if A' and x are defined as in the proof of Theorem 3, then $A'[x]$ is an integral domain. Hence it follows that if A is an integral domain, and x is an indeterminate over A, then $A[x]$ is an integral domain.

2. Modify Definition 1 for the case that the subring A is not assumed to be commutative, but where it is assumed that $at = ta$, for each $a \in A$. Carry out proofs of Theorems 1—3 for this case.

4.   *Polynomial Domains Over a Field*

Suppose t is an indeterminate over the (commutative) ring A and $p \in A[t]$, $p \neq 0$. By Definition 2, p has a nonzero coefficient. Hence there is a largest integer $m \in N$ such that

$$p = \sum_{i=0}^{m} a_i t^i,$$

where $a_m \neq 0$. By Theorem 1, the integer m and the coefficients $a_0, a_1, \ldots, a_m$ are completely determined by p.

DEFINITION 3.   The largest integer m such that $a_m \neq 0$ is the *degree* of p; it is denoted by "∂p," so that if m is the degree of p, then

$$\partial p = m.$$

The coefficient a_m is the *leading* coefficient of p; p is a *monic* polynomial if and only if $a_m = 1$. The zero polynomial is assigned no degree. A polynomial p is a *constant* polynomial if and only if $p = a_0$ where $a_0 \in A$. (*Note:* $a_0 = 0$ is admitted.) Let $p = \sum\limits_{i=0}^{m} a_i t^i$ and $a_j \neq 0$; then the *degree of the term* $a_j t^j$ is j.

Exercise. Let t be an indeterminate over A. Prove: (a) If $p, q \in A[t]$, then $\partial(pq) \leq \partial p + \partial q$; (b) if $q \neq -p$, then $\partial(p + q) \leq \text{Max}(\partial p, \partial q)$; (c) if A is an integral domain, $\partial(pq) = \partial p + \partial q$.

THEOREM 4.   If t is an indeterminate over a field F, then for each a and each b in $F[t]$, $b \neq 0$, there is a unique pair $q, r \in F[t]$ such that

(11) $$a = qb + r,$$

where $r = 0$ or $0 \leq \partial r < \partial b$ (q is the *quotient* and r is the *remainder*).

Remark: In the example following the proof we shall see that the proof of Theorem 4 carries through, in formalized style, the process of division by polynomials familiar to us from elementary algebra.

Proof: Existence. Let $a = \sum\limits_{i=0}^{m} a_i t^i$, $b = \sum\limits_{j=0}^{n} b_j t^j$, $b_n \neq 0$. If $n > m$ or if $a = 0$, the existence part of the theorem is established by noting that $q = 0$ and $r = a$ do the trick. Henceforth we assume that $m \geq n$, $a_m \neq 0$, and proceed by induction on $m = \partial a$.

   If $m = 0$, then also $n = 0$. In this case $a = a_0$, and $b = b_0$. Hence $q = a_0 b_0^{-1}$ and $r = 0$ are desired polynomials.

   Let $m > 0$ and assume that the result holds for all polynomials of degree smaller than m. Set

$$a' = a - a_m b_n^{-1} t^{m-n} b;$$

clearly, $\partial a' < m$, and therefore by the induction hypothesis there exist $q', r \in F[t]$ such that

$$a' = q'b + r,$$

where $r = 0$ or $0 \leq \partial r < \partial b$. Hence

$$a - a_m b_n^{-1} t^{m-n} b = q'b + r,$$

or,

$$a = (a_m b_n^{-1} t^{m-n} + q')b + r$$
$$= qb + r,$$

where $q = a_m b_n^{-1} t^{m-n} + q'$. This completes the existence part of the proof.

Uniqueness. Suppose $q_i, r_i, i = 1, 2$ are polynomials in $F[t]$ satisfying (11). Then

$$q_1 b + r_1 = q_2 b + r_2,$$

whence

$$(q_1 - q_2)b = r_2 - r_1.$$

If $q_1 - q_2 \neq 0$, then $r_2 - r_1 \neq 0$ and by the Exercise of page 212, $\partial((q_1 - q_2)b) \geq \partial b$ and $\partial(r_2 - r_1) < \partial b$, a contradiction. Therefore $q_1 = q_2$ and $r_1 = r_2$.

Q.E.D.

EXAMPLE 4. Let $a, b \in Q[t]$ where Q is the field of rational numbers and

$$a = t^5 - t^4 - 5t^3 + 8t^2 + 1,$$

and

$$b = t^2 - 3t + 2;$$

thus $m = 5$, $a_5 = 1$ and $n = 2$, $b_2 = 1$. Then

$$\begin{aligned}
a' &= a - 1 \cdot 1^{-1} \cdot t^{5-2} \cdot (t^2 - 3t + 2) \\
&= t^5 - t^4 + 8t^2 + 1 - t^5 - 3t^4 - 2t^3 \\
&= 2t^4 - 7t^3 + 8t^2 + 1,
\end{aligned}$$

where $\partial a' = 4$ and $a'_4 = 2$. Repeating the process with a' and b, we obtain

$$\begin{aligned}
a'' &= a' - 2 \cdot 1^{-1} \cdot t^{4-2} \cdot (t^2 - 3t + 2) \\
&= -t^3 + 4t^2 + 1,
\end{aligned}$$

where $\partial a'' = 3$ and $a''_3 = -1$. Using the polynomials a'' and b, our process yields

$$\begin{aligned}
a''' &= a'' - (-1) \cdot 1^{-1} \cdot t^{3-2} \cdot (t^2 - 3t + 2) \\
&= t^2 + 2t + 1,
\end{aligned}$$

where $\partial a''' = 2$ and $a'''_2 = 1$. Finally

$$\begin{aligned}
a^{iv} &= a''' - 1 \cdot 1^{-1} \cdot (t^2 - 3t + 2) \\
&= 5t - 1.
\end{aligned}$$

From the foregoing we have

$$\begin{aligned}
a &= a' + t^3(t^2 - 3t + 2), \\
a' &= a'' + 2t^2(t^2 - 3t + 2), \\
a'' &= a''' + (-1)t(t^2 - 3t + 2), \\
a''' &= a^{iv} + (t^2 - 3t + 2),
\end{aligned}$$

whence, by substitution, we find

$$a = (t^3 + 2t^2 - t + 1)(t^2 - 3t + 2) + (5t - 1).$$

Thus the quotient and remainder are $t^3 + 2t^2 - t + 1$ and $5t - 1$, respectively.

Of course, in practice one never carries out a division of one polynomial by another by following the method of the theorem, in detail, as above. One usually resorts to the short cut, long-division, familiar to all. Thus:

$$
\begin{array}{r}
t^3 + 2t^2 - t + 1 \qquad \text{(= quotient)} \\
(b =) \quad t^2 - 3t + 2 \overline{\bigm)\; t^5 - t^4 - 5t^3 + 8t^2 + 1 \quad \text{(= a)}} \\
\underline{-t^5 - 3t^4 + 2t^3 } \\
2t^4 - 7t^3 + 8t^2 \text{(= a')} \\
\underline{2t^4 - 6t^3 + 4t^2 } \\
-t^3 + 4t^2 \text{(= a'')} \\
\underline{-t^3 + 3t^2 - 2t } \\
t^2 + 2t + 1 \quad \text{(= a''')} \\
\underline{t^2 - 3t + 2 } \\
5t - 1 \quad \text{(= remainder).}
\end{array}
$$

The reader should compare the long-division with the step-by-step illustration of Theorem 4.

Theorem 4 provides a division algorithm, for polynomial domains over fields, which is quite similar to the division algorithm for the integers. The latter algorithm was used to prove that Z is a PID. Hence, by analogy, we conjecture that the division algorithm for polynomials over a field F can be used to prove

THEOREM 5. $F[t]$ is a PID.

   Proof: Let I be an ideal in $F[t]$; we show that I is generated by a single polynomial. If $I = \{0\}$, then the zero polynomial generates I. If $I \neq \{0\}$, then I contains a polynomial $b \neq 0$ of smallest degree, say $\partial b = m \geq 0$ (details? Note that the WOP is used here.) We claim that b generates I.

   Since $s \cdot b \in I$ for each $s \in F[t]$ (why?), it suffices to show that for each $a \in I$, there is a polynomial q in $F[t]$ such that $a = qb$. By Theorem 4, there exist q and r in $F[t]$ such that

$$a = qb + r,$$

where $r = 0$, or else $0 \leq \partial r < \partial b$. Now $r = a - qb \in I$. Hence if $r \neq 0$, then r is a polynomial in I of degree less than that of b. This is a contradiction, since b was chosen as a polynomial of smallest degree in I. Therefore $r = 0$ and so $a = qb$.

                                                                                           Q.E.D.

COROLLARY. $F[t]$ is a UFD.

   Proof: Theorem 35, Chapter 4, page 190.

                                                                                           Q.E.D.

Not only is $F[t]$ a UFD, but one can show that for each field F, $F[t]$ contains prime elements (see Exercise 7, page 215, below). The prime elements in $F[t]$ are usually called "irreducible" polynomials; those which are not irreducible are said to be "reducible."

Since $F[t]$ is a PID with unity element, we may list several important properties of $F[t]$ which are immediate consequences of results established in Chapter 4.

(1) If a, b are polynomials in $F[t]$ such that $a \neq 0$ or $b \neq 0$, then a and b have a greatest common divisor which is unique to within unit factors (Theorem 33, Chapter 4, page 184); by Exercise 6, below, the unit factors are the nonzero constant polynomials.

(2) If (a, b) is the g.c.d. of a and b, then there exist polynomials r, $s \in F[t]$ such that

$$(a, b) = ra + sb.$$

(3) If $p, a, b \in F[t]$ where p is irreducible, and if $p \mid ab$, then $p \mid a$ or $p \mid b$. More generally, if $(p, a) = 1$, then $p \mid ab$ implies $p \mid b$.

(4) The unique factorization property of $F[t]$ is expressed thus: If $p \in F[t]$ is a polynomial of positive degree, then there are irreducible polynomials $p_1, p_2, \ldots, p_k$, $(k \geq 1)$, and positive integers $n_1, n_2, \ldots, n_k$ and a nonzero element $c \in F$ such that

$$p = c p_1^{n_1} p_2^{n_2} \cdots p_k^{n_k}.$$

This factorization of p is unique to within units and the order of the factors.

Exercises

1. Let Q be the field of rational numbers, x an indeterminate over Q. Find q and r in $Q[x]$ in case (i) $a = 2x^4 + 3x^3 + 6x^2 + 4x + 6$ and $b = 4 + 3x + 2x^2$; (ii) $a = x^4 + 3x^3 - 2x^2 + x + 1$ and $b = x^2 + 3x$.

2. Let $a = x^4 + 3x^3 - 2x^2 + x + 1$, $b = x^3 + 3x$ be elements in $Z_n[x]$, x an indeterminate over Z_n. Find q and r in case (i) $n = 5$; (ii) $n = 7$; (iii) $n = 11$.

3. By repeated use of the division algorithm, find g.c.d.'s for the several pairs of polynomials in Exercises 1 and 2.

4. Is $2x^3 + 3x^2 + 1$ an irreducible polynomial in $Z_n[x]$ where (i) $n = 5$; (ii) $n = 7$?

5. Decompose $x^4 - 3x^2 - 4 \in F[x]$ into a product of irreducible polynomials where (i) $F = Q$ the field of rational numbers; (ii) $F = Q[i] = \{a + bi \mid a, b \in Q$ and $i^2 = -1\}$.

6. Prove: An element $p \in F[x]$, $p \neq 0$, is a unit if and only if $\partial p = 0$, hence if and only if $p = a_0$ where a_0 is a nonzero element in the field F.

7. Prove that every linear polynomial (i.e., polynomial of degree one) in $F[x]$ is irreducible.

8. Prove that $Z_2[x]$ contains monic, irreducible polynomials of degree 2, i.e., monic, irreducible, quadratic polynomials. (*Hint:* Show that the number of monic, quadratic polynomials exceeds the number of monic, quadratic, reducible polynomials.)

9. Prove: If F is a finite field, then $F[x]$ contains monic, irreducible, quadratic polynomials.

10. Let F be a field and x be an indeterminate over F. Prove: (i) If $p_1, p_2, \ldots, p_n$ are nonzero polynomials in $F[x]$, then these polynomials have a g.c.d. which is unique to within unit factors. (ii) If d is the g.c.d. of $p_1, p_2, \ldots, p_n$, then there exist $q_1, q_2, \ldots, q_n$ in $F[x]$ such that

$$d = p_1 q_1 + p_2 q_2 + \ldots + p_n q_n.$$

5. Unique Factorization in Polynomial Domains

In Section 4 we saw that if F is a field and x is an indeterminate over F, then $F[x]$ is a PID; consequently, $F[x]$ is a UFD. The purpose of this section is to establish the following more general result:

THEOREM 6. *If the integral domain D is a UFD, so is the polynomial domain $D[x]$, where x is an indeterminate over D.*

Let Q be the quotient field of D, and let $Q[y]$ be the polynomial domain in the indeterminate y over Q. The set

$$S = \left\{ \sum_{i=0}^{m} a_i y^i \,\middle|\, m \in N \text{ and } a_i \in D, i = 0, 1, 2, \ldots, m \right\}$$

is evidently a subdomain of $Q[y]$. In fact, $S = D[y]$ and by Exercise 4, page 206, y is an indeterminate over D. By Theorem 2, $D[x]$ and $D[y]$ are isomorphic. Therefore if we can prove that $D[y]$ is a UFD, the same must be true of $D[x]$. Thus it suffices to prove Theorem 6 for $D[y]$.

To show that $D[y]$ is a UFD, we shall prove:

If $p \in D[y]$, then there is a unit ϵ in D and irreducible polynomials $p_1, p_2, \ldots, p_n$ in $D[y]$ such that

$$(12) \qquad\qquad p = \epsilon p_1 p_2 \cdots p_n.$$

Moreover, the factorization (12) is unique to within units and the order of the factors. Now since $D[y] \subset Q[y]$, $p \in D[y]$ implies $p \in Q[y]$. But *we already know that $Q[y]$ is a UFD.* Hence if $p \in D[y]$, then

$$(13) \qquad\qquad p = c q_1 q_2 \cdots q_n,$$

where $c \in Q$ and $q_1, q_2, \ldots, q_n$ are irreducible polynomials in $Q[y]$. In the reasoning below we shall use the available factorization (13) in order to obtain the desired one, (12). Thus the proof that $D[y]$ is a UFD relies heavily upon the fact (Corollary to Theorem 5) that $Q[y]$ is a UFD, and it contains $D[y]$ as a subdomain.

To simplify the notation, we shall adopt the convention that

$$p, p', p_1, p_2, \ldots, p'_1, p'_2, \ldots, \text{etc.,}$$

are variables with range $D[y]$, and that

$$q, q', q_1, q_2, \ldots, q'_1, q'_2, \ldots, \text{etc.,}$$

are variables on $Q[y]$.

Since D is a UFD, every set $\{c_1, c_2, \ldots, c_k \,|\, k \geq 2, c_i \in D, 1 \leq i \leq k,$ c_i not all zero$\}$ has a g.c.d. which is unique to within unit multiples.

DEFINITION 4. A polynomial $p = \sum_{i=0}^{m} a_i y^i \in D[y]$ is *primitive* if and only if the g.c.d. of $\{a_0, a_1, \ldots, a_m\}$ is 1.

Exercises

1. Prove: The only units in $D[y]$ are the units in D.

2. Prove: If $p \in D[y]$ is an irreducible polynomial, then either $p = a$ where $a \in D$ is prime, or else p is a primitive polynomial.

3. Is $3y + 6$ reducible in $Q[y]$? In $Z[y]$?

4. Is a primitive polynomial irreducible? If not, give a counterexample.

Our proof of Theorem 6 requires a sequence of lemmas, the first being a famous result due to Gauss.

LEMMA 1. If $p = \sum_{i=0}^{m} a_i y^i$ and $p_1 = \sum_{j=0}^{n} b_j y^j$ are primitive polynomials in $D[y]$ such that $a_m b_n \neq 0$, then their product is primitive.

Proof: Let

$$pp' = \sum_{k=0}^{m+n} c_k y^k,$$

where

$$c_0 = a_0 b_0, c_1 = a_0 b_1 + a_1 b_0, \ldots, c_{m+n} = a_m b_n,$$

and suppose pp' is not primitive. Then there. is a prime element $d \in D$ such that $d \,|\, c_i$, $0 \leq i \leq m + n$. However, since p and p' are primitive, some coefficient of p and some coefficient of p' is not divisible by d. Consequently, there exist smallest integers r and s, $0 \leq r \leq m$, $0 \leq s \leq n$, such that $d \nmid a_r$ and $d \nmid b_s$. Thus

$$d \,|\, a_i, \ 0 \leq i < r \text{ and } d \,|\, b_j, \ 0 \leq j < s,$$

and

$$d \nmid a_r \text{ and } d \nmid b_s.$$

Now

$$c_{r+s} = a_0 b_{r+s} + a_1 b_{r+s-1} + \cdots + a_r b_s + a_{r+1} b_{s-1}$$
$$+ \cdots + a_{r+s} b_0,$$

and since $d \,|\, c_{r+s}$ and $d \,|\, a_i b_{r+s-i}$, $i \neq r$, it follows that $d \,|\, a_r b_s$. But since d is prime, $d \,|\, a_r$ or $d \,|\, b_s$, a contradiction. Hence pp' is primitive.

Q.E.D.

LEMMA 2. Let p, p' be primitive polynomials in $D[y]$, and let $e, f \in Q$ where

$$ep = fp'.$$

Then $p = \epsilon p'$ and $e = \eta f$ where ϵ and η are units *in* D.

Proof: First consider the case that $e = 1$. Since $f \in Q$, $f = a/b$ where $a, b \in D$; we may assume that g.c.d. $(a, b) = 1$. Hence

$$p = \frac{a}{b} p',$$

and so

$$bp = ap'.$$

Since g.c.d. $(a, b) = 1$ and since $a \,|\, bp$, we deduce $a \,|\, p$. Hence $a \,|\, a_i$, $0 \leq i \leq m$. But since p is primitive, a must be a unit in D. A similar argument shows that b is also a unit in D. Therefore $f = a/b$ is a unit in D, and the lemma is proved in this case.

Now suppose $e = a/b$, $f = c/d$ where $a, b, c, d \in D$; we may assume that

g.c.d. (a, b) = g.c.d. (c, d) = g.c.d. (a, c) = g.c.d. (b, d) = 1.

Then

$$\frac{a}{b} p = \frac{c}{d} p'$$

implies

$$p = \frac{bc}{ad} p',$$

where g.c.d. $(ad, bc) = 1$. By the first part of the proof, ad and bc are units in D, hence so are a, b, c and d. Consequently $p = \epsilon p'$ where $\epsilon = bc/ad$ is a unit in D and $c/d = \epsilon(a/b)$.

Q.E.D.

LEMMA 3. Let q be a polynomial in $Q[y]$. Then there exists a primitive polynomial p *in* $D[y]$ such that

$$q = cp,$$

where $c \in Q$. The element c and the primitive polynomial p are determined to within units in D. Moreover, if $q \in D[y]$, then $c \in D$.

Proof: Let

$$q = \sum_{i=0}^{m} \frac{c_i}{b_i} y^i,$$

where $c_i, b_i \in D$, $0 \le i \le m$, $c_m \ne 0$, and set $b = b_0 b_1 \ldots b_m$. Since D is a UFD,

$$q = \frac{1}{b}(c_0 b_1 \cdots b_m + b_0 c_1 b_2 \cdots b_m y + \cdots + b_0 b_1 \cdots b_{m-1} c_m y^m)$$

$$= \frac{a}{b}(a_0 + a_1 y + \cdots + a_m y^m)$$

$$= \frac{a}{b} p,$$

where $a = $ g.c.d. $(c_0 b_1 \cdots b_m, b_0 c_1 b_2 \cdots b_m, \ldots, b_0 b_1 \cdots b_{m-1} c_m)$, and so $p = \sum_{i=0}^{m} a_i y^i \in D[y]$ is primitive. Lemma 2 shows that a/b and q are determined to within units in D. Clearly, if $q \in D[y]$, then $b = 1$ and so $a = a/1 \in D$.

Q.E.D.

LEMMA 4. Let $q, q_1, q_2, \ldots, q_n$, $n \ge 2$, be polynomials in $Q[y]$ such that $q = q_1 q_2 \cdots q_n$. Suppose that

$$q = cp \text{ and } q_i = c_i p_i, \; 0 \le i \le n,$$

where $c, c_i \in Q$ and p, p_i are primitive polynomials in $D[y]$. Then

$$p = \epsilon p_1 p_2 \cdots p_n,$$

and

$$c = \eta c_1 c_2 \cdots c_n,$$

where ϵ and η are units in D.

Proof: By induction. If $n = 2$, then

$$cp = (c_1 c_2) p_1 p_2,$$

and by Lemma 1, $p_1 p_2$ is a primitive polynomial in $D[y]$. By Lemma 3, $p = \epsilon p_1 p_2$ and $c = \eta c_1 c_2$ where ϵ and η are units in D. The completion of the induction is left to the reader.

Q.E.D.

LEMMA 5. Let p be a primitive polynomial in $D[y]$. Then p is a product of nonconstant polynomials in $D[y]$ if and only if p is reducible in $Q[y]$.

Proof: First suppose $p = p_1 p_2$ where p_1 and p_2 are nonconstant polynomials in $D[y]$. Then p_1 and p_2 are both polynomials of positive degree in $Q[y]$, and so neither p_1 nor p_2 is a unit in $Q[y]$. Hence p is reducible in $Q[y]$.

Conversely, suppose p is reducible in $Q[y]$, say, $p = q_1 q_2$

where $q_1, q_2 \in Q[y]$ and $\partial q_1 > 0$, $\partial q_2 > 0$. By Lemma 3 there exist elements c and d in Q and primitive polynomials p_1 and p_2 in $D[y]$ such that

$$q_1 = cp_1 \text{ and } q_2 = dp_2.$$

Then

$$p = (cd)p_1p_2,$$

and by Lemma 1, p_1p_2 is a primitive polynomial in $D[y]$. By Lemma 2 there exists a unit ϵ in D such that $p = \epsilon p_1 p_2$ and therefore p is a product of the nonconstant polynomials ϵp_1 and p_2 in $D[y]$.

Q.E.D.

Proof of Theorem 6: Let $p \in D[y]$; we show first that

(14) $$p = \epsilon d_1^{m_1} d_2^{m_2} \cdots d_s^{m_s} p_1^{n_1} p_2^{n_2} \cdots p_t^{n_t},$$

where the d's are primes in D, the p's are irreducible polynomials in $D[y]$, ϵ is a unit in D, and the m's and n's are positive integers.

Since $p \in D[y]$, also $p \in Q[y]$. Since the latter is a UFD whose only units are the nonzero elements in Q,

$$p = eq_1^{n_1} q_2^{n_2} \cdots q_t^{n_t},$$

where $e \in Q$, the q_i are irreducible polynomials in $Q[y]$, $1 \le i \le t$, and the n's are positive integers. By Lemma 3,

$$p = cp' \text{ and } q_i = c_ip_i', \quad 1 \le i \le t,$$

where $c, c_i \in Q$ and p' and the p_i' are primitive polynomials in $D[y]$. Each p_i' is irreducible in $D[y]$, since otherwise the corresponding q_i would be reducible. Thus

$$cp' = (ec_1c_2 \cdots c_t)p_1'^{n_1} p_2'^{n_2} \cdots p_t'^{n_t},$$

whence by Lemma 4,

$$p' = \tau p_1'^{n_1} p_2'^{n_2} \cdots p_t'^{n_t},$$

and

$$c = \eta(ec_1c_2 \cdots c_t),$$

where τ and η are units in D. By Lemma 3, $c \in D$ and so $ec_1c_2 \cdots c_t \in D$. Since D is a UFD,

$$c = \eta(ec_1c_2 \cdots c_t) = (\eta\mu)d_1^{m_1} d_2^{m_2} \cdots d_s^{m_s},$$

where the d's are primes in D, η as well as $\eta\mu$ are units in D, and the m's are positive integers. Thus we have proved that equation (14) holds.

To establish the uniqueness of factorization in $D[y]$ suppose that

$$d_1^{m_1} d_2^{m_2} \cdots d_s^{m_s} p_1^{n_1} p_2^{n_2} \cdots p_t^{n_t} = f_1^{j_1} f_2^{j_2} \cdots f_u^{j_u} p_1'^{k_1} p_2'^{k_2} \cdots p_v'^{k_v},$$

where the d's and f's are primes in D, the p's and p''s are primitive polynomials in $D[y]$ no one of which is a product of polynomials of positive degree, and the m's, n's, j's and k's are positive integers. By Lemma 2,

(15) $$d_1^{m_1} d_2^{m_2} \cdots d_s^{m_s} = \epsilon f_1^{j_1} f_2^{j_2} \cdots f_u^{j_u},$$

and

(16) $$p_1^{n_1} p_2^{n_2} \cdots p_t^{n_t} = \eta p_1'^{k_1} p_2'^{k_2} \cdots p_v'^{k_v},$$

where ϵ and η are units in D. Since D is a UFD, and since the d's and f's are primes in D, the two factorizations in (15) are the same up to unit factors and to within orders of the factorizations. Further, the factorizations (14) may be regarded as factorizations in $Q[y]$. By Lemma 5, each p and each p' is irreducible *in* $Q[y]$. Since $Q[y]$ is a UFD, the factorizations in (16) are the same to within unit factors and to within orders of the factorizations.

Q.E.D.

In Section 12, Chapter 4, we promised an example of a UFD which is not a PID. The results of this section will be employed to provide such an illustration.

EXAMPLE 5. Let x be an indeterminate over the domain Z of integers. Since Z is a UFD, so is $Z[x]$ (Theorem 6). We exhibit an ideal in $Z[x]$ which is not principal, whence it follows that $Z[x]$ is not a PID.

Let $I = 2Z[x] + xZ[x]$ be the ideal generated by the set $\{2, x\}$. If I were principal, say, with generator p, then since $Z[x]$ has a unity element, $I = pZ[x]$. Now, since $2, x \in I, p \mid 2$ and $p \mid x$. But 2 and x are both primes in $Z[x]$ (why?), and therefore p must be a unit. Hence $I = pZ[x] = Z[x]$. Since $1 \in I$, there exist $a, b \in Z[x]$ such that

(17) $$1 = 2a + xb.$$

If $a \neq 0$, then $2a + xb$ is a polynomial having constant term divisible by 2. Equation (17) shows that this is impossible. If $a = 0$, then $2a + xb$ has zero as its constant term, and this too is impossible. Thus in every case the assumption that I is principal yields a contradiction. Therefore $Z[x]$ is not a PID.

Also, on page 194, we claimed that there exist prime ideals which are not maximal. An example of such an ideal can be found in the ring $Z[x]$.

EXAMPLE 6. Let (x) be the principal ideal in $Z[x]$ generated by the element x. Since $Z[x]$ is a ring with unity element, we know that

$$(x) = xZ[x],$$

hence (x) consists of all polynomials in $Z[x]$ which are divisible by x. Now if $p, q \in Z[x]$ and

$$pq \in (x),$$

then $x \mid pq$. But since x is a prime and $Z[x]$ is a UFD, $x \mid pq$ implies $x \mid p$ or $x \mid q$, i.e., $p \in (x)$ or $q \in (x)$. Thus (x) is a prime ideal in $Z[x]$.

We now show that (x) is not maximal. Indeed, the ideal I of Example 5, generated by $\{2, x\}$, clearly contains (x). Since $2 \in I$ but $2 \notin (x)$,

$$(x) \subsetneqq I.$$

Further, as in Example 5, $1 \notin I$, so that

$$I \subsetneqq Z[x].$$

Consequently

$$(x) \subsetneqq I \subsetneqq Z[x]$$

and therefore (x) is not maximal.

6. Polynomial Rings in Two Indeterminates

Let A be a ring with unity element and let x be an indeterminate over A. Then $A[x]$ is again a ring with unity element, and by Theorem 3 there exists an indeterminate y over $A[x]$. We shall set

(18) $A[x, y] = A[x][y];$

the elements of $A[x, y]$ are polynomials

$$p = \alpha_0 + \alpha_1 y + \alpha_2 y^2 + \cdots + \alpha_n y^n,$$

where each $\alpha_i \in A[x]$, $0 \le i \le n$, hence

$$\alpha_i = a_{0i} + a_{1i}x + a_{2i}x^2 + \cdots + a_{r_i i}x^{r_i}, \quad 0 \le i \le n.$$

Consequently, if $p \in A[x, y]$, then

(19) $$p = \sum_{i_2=0}^{n_2} \sum_{i_1=0}^{n_1} a_{i_1 i_2} x^{i_1} y^{i_2},$$

where n_1 and n_2 are nonnegative integers and the $a_{i_1 i_2}$ are elements in A.

Now if $p = 0$, then since y is an indeterminate over $A[x]$ it follows that each $\alpha_i = 0$, $0 \le i \le n$. On the other hand, since $\alpha_i \in A[x]$ and since x is an indeterminate over A, from $\alpha_i = 0$ it follows that each $a_{i_1 i_2} = 0$. In short:

If $p = 0$, then each coefficient $a_{i_1 i_2} = 0$.

Conversely, if each of the coefficients $a_{i_1 i_2} = 0$ (in (19)), then it is clear that $p = 0$.

Further, let $q \in A[x, y]$, say,

$$q = \sum_{i_2=0}^{m_2} \sum_{i_1=0}^{m_1} b_{i_1 i_2} x^{i_1} y^{i_2},$$

and suppose that the coefficient $a_{n_1 n_2}$ of p in (19) is nonzero and that $p = q$.

Then from the foregoing reasoning, we deduce $m_1 = n_1$, $m_2 = n_2$ and $b_{i_1 i_2} = a_{i_1 i_2}$ for all the ordered pairs (i_1, i_2) (details?). These results justify the use of the term "degree" as we now define it.

> **DEFINITION 5.** Let $a_{j_1 j_2} x^{j_1} y^{j_2}$ be a term in (19) such that $a_{j_1 j_2} \neq 0$. Then $j_1 + j_2$ is the *degree* of the *term* $a_{j_1 j_2} x^{j_1} y^{j_2}$. The *degree* of the *polynomial* p is the maximum of the sums $j_1 + j_2$ for which the coefficients $a_{j_1 j_2}$ are non-zero. A polynomial p is a *constant polynomial* if and only if it has no term of positive degree. The degree of a polynomial is denoted by "∂p."

Exercises

1. Prove: If D is an integral domain, so is $D[x, y]$.
2. Let $p, q \in A[x, y]$. Prove that $\partial(pq) \leq \partial p + \partial q$ and if $q \neq -p$, then $\partial(p + q) \leq \text{Max}\,(\partial p, \partial q)$. Show that if A is an integral domain, then $\partial(pq) = \partial p + \partial q$.

The concept of polynomial ring in two indeterminates can, of course, be extended to polynomial rings in n indeterminates. However, as we are not going to develop the theory of such polynomial rings, we shall drop this subject forthwith.

We conclude Section 6 with another example of a UFD which is not a PID.

> **EXAMPLE 7.** Let F be a field, let x be an indeterminate over F, and let y be an indeterminate over $F[x]$. In accordance with (18), we write
>
> $$F[x, y] = F[x][y].$$
>
> By Theorem 5, $F[x]$ is a PID and therefore (Corollary to Theorem 5) it is also a UFD. By Theorem 6, taking $D = F[x]$ we see that $F[x, y] = D[y]$ is also a UFD. We now prove:
>
> $$F[x, y] \text{ is not a PID.}$$
>
> In order to prove this statement, we shall exhibit an ideal I in $F[x, y]$ which is not principal. Let I be the ideal generated by the set $\{x, y\}$. Since $F[x, y]$ has a unity element, 1,
>
> $$I = \{fx + gy \,|\, f, g \in F[x, y]\}$$
>
> (see Exercise 3, page 171).
>
> Since x, y are both elements in I, it follows that I is not the zero ideal. Further, it is not the unit ideal (i.e., the whole ring). For, in the contrary case, $1 \in I$, and so there exist $p, q \in F[x, y]$ such that
>
> $$1 = px + qy.$$

This is a contradiction, since $px + qy$ has no term of degree zero, whereas 1 is a polynomial of degree zero. Therefore I is not the unit ideal.

If $a \in F$, $a \neq 0$, then $a \notin I$. For, if $a \in I$, then since $a \neq 0$, a^{-1} is a constant polynomial and so $1 = a^{-1}a \in I$, a contradiction.

Now suppose I is a principal ideal generated by the polynomial $p \in F[x, y]$. The above argument shows that p is not a constant polynomial, hence $\partial p \geq 1$. However, since $x \in I$ and $y \in I$, there exist polynomials f, g such that

$$x = fp \text{ and } y = gp.$$

Since $\partial x = \partial y = 1$, Exercise 2, page 223, shows that $\partial p = 1$ and $\partial f = \partial g = 0$. But since $\partial p = 1$,

$$p = a_0 + a_1 x + a_2 y,$$

where $a_1 \neq 0$ or $a_2 \neq 0$. The fact that $\partial f = \partial g = 0$ implies that $f = b$ and $g = c$ where b and c are nonzero elements in F. Thus,

$$x = b(a_0 + a_1 x + a_2 y),$$

whence $a_0 = a_2 = 0$ and $a_1 \neq 0$. On the other hand,

$$y = c(a_0 + a_1 x + a_2 y)$$

implies $a_2 \neq 0$ and $a_1 = 0$, in contradiction of the preceding result.

Thus, the assumption that I is principal yields a contradiction, and so $F[x, y]$ is not a PID.

7. Polynomial Functions and Polynomials

In this section we propose to show that the concept of polynomial in the sense of Definition 1 is a generalization of the elementary concept of polynomial. Let us recall that the polynomials of elementary mathematics are *functions*, usually defined on subsets of the complex numbers. Thus, if A is a subring of the complex numbers C then a *polynomial function on A* is a function $f: A \longrightarrow A$ defined by

$$(20) \qquad f(x) = a_0 + a_1 x + \cdots + a_m x^m, \quad x \in A,$$

where $m \in N$ and the coefficients $a_0, a_1, \ldots, a_m$ are elements in A. We shall prove that from the proper viewpoint, the polynomial functions (20) are polynomials in an element t over a ring A^* to be specified below; moreover, t will turn out to be an indeterminate over A^*.

Before proceeding to the main discussion, we offer an example which may aid the reader in understanding the difference between polynomial functions and polynomials, in the sense of Definition 1.

EXAMPLE 8. Let $A = Q$ be the field of rationals, and let $B = R$ be the real field. For each element $t \in R$,

$$2 + 3t - t^2 + \frac{1}{2}t^3$$

is a polynomial in t over Q. Specifically,

$$2 + 3\pi - \pi^2 + \frac{1}{2}\pi^3,$$

$$2 + 3(\sqrt{2}) - (\sqrt{2})^2 + \frac{1}{2}(\sqrt{2})^3,$$

$$2 + 3e - e^2 + \frac{1}{2}e^3,$$

$$2 + 3\left(-\frac{1}{7}\right) - \left(-\frac{1}{7}\right)^2 + \frac{1}{2}\left(-\frac{1}{7}\right)^3,$$

are polynomials in π, $\sqrt{2}$, e and $-\frac{1}{7}$, respectively, over Q. There are infinitely (in fact, uncountably) many polynomials over Q having the coefficients $a_0 = 2$, $a_1 = 3$, $a_2 = -1$ and $a_3 = \frac{1}{2}$. On the other hand, there is exactly one polynomial function,

$$f(x) = 2 + 3x - x^2 + \frac{1}{2}x^3, \quad x \in Q,$$

with the same coefficients.

To carry out the program of this section, let A be a subfield of C (= complex field) and let $\mathcal{F}(A)$ be the set of all functions $f: A \longrightarrow A$. In the usual elementary cases, A is the field of rationals, or reals, or complex numbers. If $f, g \in \mathcal{F}(A)$, define sum and product on $\mathcal{F}(A)$ by

$$(f + g)(x) = f(x) + g(x), \quad x \in A,$$

and

$$(fg)(x) = f(x)g(x), \quad x \in A,$$

respectively. (Note that the $+$ and $\cdot$ on the right-hand sides are the sum and product in A. The $+$ and $\cdot$ on the left-hand sides are the sum and product being defined in $\mathcal{F}(A)$). Among the functions in $\mathcal{F}(A)$ are the zero function,

$$c_0(x) = 0, \quad x \in A,$$

and the unity function,

$$c_1(x) = 1, \quad x \in A,$$

where 1 is the unity element of A.

THEOREM 7. $\mathcal{F}(A)$ is a commutative ring having c_0 as zero element and c_1 as unity element.

Proof: We verify a few of the conditions and leave the rest to the reader.

First, c_0 is the zero element of $\mathscr{F}(A)$. For, if $f \in \mathscr{F}(A)$, then for each $x \in A$,

$$\begin{aligned}
(c_0 + f)(x) &= c_0(x) + f(x) \\
&= 0 + f(x) \\
&= f(x),
\end{aligned}$$

so that $c_0 + f = f$. Next, if one defines $-f$ by

$$(-f)(x) = -(f(x)), \quad x \in A,$$

then for each $x \in A$,

$$((-f) + f)(x) = -(f(x)) + f(x) = 0 = c_0(x).$$

Therefore

$$(-f) + f = c_0,$$

and so $-f$ is the additive inverse of f.

Let f, g and h be elements in $\mathscr{F}(A)$. Then, if $x \in A$,

$$\begin{aligned}
(f(g + h))(x) &= f(x)(g + h)(x) \\
&= f(x)(g(x) + h(x)) \\
&= f(x)g(x) + f(x)h(x) \\
&= (fg)(x) + (fh)(x) \\
&= (fg + fh)(x),
\end{aligned}$$

whence

$$f(g + h) = fg + fh.$$

Etc.

Q.E.D.

The zero element c_0 and the unity element c_1 are special instances of *constant functions*.

DEFINITION 6. · Let $a \in A$; the function $c_a \in \mathscr{F}(A)$ defined by

$$c_a(x) = a, \quad x \in A,$$

is the *constant function determined by a*.

In particular, c_0 and c_1 are the constant functions determined by 0 and 1, respectively.

Let $\mathscr{CF}(A)$ be the set of all constant functions on A. Obviously, $\mathscr{CF}(A) \subset \mathscr{F}(A)$. But, in addition, for each $x \in A$,

$$\begin{aligned}
(c_a - c_b)(x) &= c_a(x) - c_b(x) \\
&= a - b \\
&= c_{a-b}(x),
\end{aligned}$$

so that $c_a - c_b = c_{a-b}$ is an element in $\mathscr{CF}(A)$. And similarly we see that $c_a c_b = c_{ab} \in \mathscr{CF}(A)$. Since $c_1 \in \mathscr{CF}(A)$, we have proved

THEOREM 8. $\mathscr{CF}(A)$ is a unitary subring of $\mathscr{F}(A)$.

Now, let u be the identity function on A, i.e.,

$$u(x) = x, \quad x \in A.$$

Clearly, if $m \in N$,

$$u^m(x) = x^m, \quad x \in A,$$

and if c_a is the constant function determined by a,

$$(c_a u^m)(x) = c_a(x)u^m(x)$$
$$= ax^m, \quad x \in A.$$

Further, one verifies easily that

$$(c_a u^m + c_b u^n)(x) = ax^m + bx^n, \quad x \in A.$$

Hence, if

$$f(x) = a_0 + a_1 x + \cdots + a_m x^m, \quad x \in A$$

is a polynomial function on A, then

$$f(x) = (c_{a_0} + c_{a_1} u + \cdots + c_{a_m} u^m)(x), \quad x \in A,$$

and therefore

$$f = c_{a_0} + c_{a_1} u + \cdots + c_{a_m} u^m.$$

Now, let $\mathscr{PF}(A)$ be the ring of all polynomial functions on A and let $A^* = \mathscr{CF}(A)$. The foregoing shows that $\mathscr{PF}(A)$ is precisely the polynomial ring

$$A^*[u] = \mathscr{CF}(A)[u].$$

Moreover, the identity function u is an indeterminate over $A^*\ (= \mathscr{CF}(A))$. Indeed, suppose $\sum_{i=0}^{m} c_{a_i} u^i$ is the zero function c_0; we show that each $c_{a_i} = c_0$, $0 \leq i \leq m$. Since $\sum_{i=0}^{m} c_{a_i} u^i = c_0$, for each $x \in A$,

$$\left(\sum_{i=0}^{m} c_{a_i} u^i \right)(x) = c_0(x) = 0,$$

or

$$a_0 + a_1 x + \cdots + a_m x^m = 0,$$

for each $x \in A$. By the elementary theory of equations (see Exercise 6 below), we know that a polynomial function is the zero function if and only if all the coefficients are zero[4]. Hence each $a_i = 0$, $0 \leq i \leq m$, and therefore each $c_{a_i} = c_0$. We have proved

---

[4] This statement, which is valid for all subfields of the complex field C, is not true for arbitrary fields. The reader may check that, in case $A = Z_2$, the assertion is false.

THEOREM 9. If

(a) A is a subfield of the field of complex numbers,
(b) $\mathscr{PF}(A)$ is the set of polynomial functions on A,
(c) $A^* = \mathscr{CF}(A)$ is the ring of constant functions on A, and
(d) u is the identity function, $u : A \longrightarrow A$,

then

$$\mathscr{PF}(A) = A^*[u],$$

and u is an indeterminate over A^*.

The exercises below are intended to review a few facts from the elementary theory of equations which motivate results to be obtained in this chapter. The following definition is required:

DEFINITION 7. Let $f \in \mathscr{PF}(A), f(x) = a_0 + a_1 x + \cdots + a_m x^m$, $x \in A$. An element $\alpha \in A$ is a *root* of f if and only if

$$f(\alpha) = a_0 + a_1 \alpha + \cdots + a_m \alpha^m = 0.$$

Exercises

1. Does $\mathscr{F}(A)$ contain zero divisors? If so, exhibit some.

2. Prove: If $a \in A$, $a \neq 0$, then the constant function c_a has no root in A.

3. Prove: An element $\alpha \in A$ is a root of $f \in \mathscr{PF}(A)$ if and only if there is an $h \in \mathscr{PF}(A)$ such that

$$f(x) = (x - \alpha)h(x), \quad x \in A,$$

where $h(x) = b_0 + b_1 x + \cdots + b_{m-1} x^{m-1}$, $x \in A$, and $b_{m-1} = a_m$.

4. Let $f, g, h \in \mathscr{PF}(A)$ and suppose $f = gh$. Prove that every root of f is a root of g or of h, and conversely.

5. Use the results of Exercises 3 and 4 to prove: If $f \in \mathscr{PF}(A), f(x) = \sum\limits_{i=0}^{m} a_i x^i$, $x \in A$, where $a_m \neq 0$, then f has at most m roots in A.

6. Prove that $f \in \mathscr{PF}(A), f(x) = \sum\limits_{i=0}^{m} a_i x^i$, $x \in A$, is the zero function in $\mathscr{PF}(A)$ if and only if

$$a_0 = a_1 = \cdots = a_m = 0.$$

7. Let $f, g \in \mathscr{PF}(A)$ where $f(x) = \sum\limits_{i=0}^{m} a_i x^i$, $x \in A$, and $g(x) = \sum\limits_{i=0}^{n} b_i x^i$, $x \in A$. Prove that $f = g$ if and only if either (i) $m \leq n$, $a_i = b_i$, $0 \leq i \leq m$, and $b_{m+1} = \cdots = b_n = 0$; or else (ii) $n < m$, $a_i = b_i$, $0 \leq i \leq n$, and $a_{n+1} = \cdots = a_m = 0$.

8. Prove that $\mathscr{PF}(A)$ is an integral domain.

9. Prove that $\mathscr{CF}(A)$ and A are isomorphic rings.

8.   Some Characterizations of Indeterminates

Let t be an indeterminate over the ring A and suppose A is a unitary subring of the ring B. (*Note:* t is in some extension ring of A, but it is immaterial whether or not $t \in B$.) If $\gamma \in B$, how is $A[t]$ related to $A[\gamma]$? We already know that if γ is also an indeterminate over A, then $A[t]$ and $A[\gamma]$ are isomorphic (Theorem 2, page 207). Explicitly, an isomorphism $\sigma_\gamma : A[t] \rightarrow A[\gamma]$ is given by

$$(21) \qquad \sigma_\gamma\left(\sum_{i=0}^{m} a_i t^i\right) = \sum_{i=0}^{m} a_i \gamma^i.$$

This isomorphism σ_γ may be thought of as a *substitution mapping determined by γ*.

Suppose we now drop the hypothesis that γ is an indeterminate. Then formula (21) still defines a substitution mapping (which may, however, not be one-one). For, if

$$\sum_{i=0}^{m} a_i t^i = \sum_{j=0}^{n} b_j t^j,$$

where $a_m \neq 0$, $b_n \neq 0$, then $m = n$ and $a_i = b_i$, $0 \leq i \leq m$. Hence

$$\sum_{i=0}^{m} a_i \gamma^i = \sum_{j=0}^{n} b_j \gamma^j,$$

i.e., $\sigma_\gamma\left(\sum_{i=0}^{m} a_i t^i\right) = \sigma_\gamma\left(\sum_{j=0}^{n} b_j t^j\right)$. Moreover, the mapping σ_γ is easily seen to have the morphism properties (details?). In short, we have proved:

If t is an indeterminate over A, then for each ring B containing A as unitary subring and for each $\gamma \in B$, $A[\gamma]$ is an epimorphic image of $A[t]$.

In the foregoing discussion, t must be an indeterminate over A, since otherwise γ will not determine a substitution mapping. To illustrate this point consider the situation of Example 2, page 204. We take $A = B = Q$ (rationals), $t = \sqrt{2}$ and $\gamma = 0$. Then in $Q[\sqrt{2}]$, the elements $2 - 1(\sqrt{2})^2$ and $6 - 1(\sqrt{2})^2 - 1(\sqrt{2})^4$ are equal; yet

$$\sigma_0(2 - 1(\sqrt{2})^2) = 2,$$

but

$$\sigma_0(6 - 1(\sqrt{2})^2 - 1(\sqrt{2})^4) = 6.$$

Consequently, the formula (21) does not define a mapping in this case.

Again, in Example 3, page 204, if $A = B = Z_3$ and $\alpha^2 = \bar{2}$, $\gamma = \bar{0}$, then

$$\bar{1} + \alpha^2 = \alpha^2 + \alpha^4,$$

yet

$$\sigma_{\bar{0}}(\bar{1} + \alpha^2) = \bar{1}$$

and

$$\sigma_{\bar{0}}(\alpha^2 + \alpha^4) = \bar{0}.$$

The foregoing suggests

THEOREM 10.   The element t is an indeterminate over A if and only if for each extension ring B of A (where A is a unitary subring) and for each $\gamma \in B$, formula (21) defines a homomorphism $\sigma_\gamma : A[t] \longrightarrow B$. In this case σ_γ is called *the substitution homomorphism determined by* γ.

Proof: We have already seen that if t is an indeterminate over A, then the substitution homomorphism $\sigma_\gamma : A[t] \longrightarrow B$ always exists.

Conversely, suppose that for each extension ring B of A and for each $\gamma \in B$, formula (21) yields a homomorphism $\sigma_\gamma : A[t] \longrightarrow B$. In particular, let us take $B = A[x]$ where x is an indeterminate over A. If we set

$$\sigma_x\left(\sum_{i=0}^{m} a_i t^i\right) = \sum_{i=0}^{m} a_i x^i,$$

then σ_x is a homomorphism. We claim that σ_x is a monomorphism. Indeed, if

$$\sum_{i=0}^{r} c_i t^i \neq \sum_{i=0}^{s} d_j t^j,$$

where, say, $r \leq s$, then either there is an integer k, $0 \leq k \leq r$, such that $c_k \neq d_k$, or else for some integer m, $r + 1 \leq m \leq s$, $d_m \neq 0$. In either case

$$\sigma_x\left(\sum_{i=0}^{r} c_i t^i\right) = \sum_{i=0}^{r} c_i x^i \neq \sum_{j=0}^{s} d_j x^j = \sigma_x\left(\sum_{j=0}^{s} d_j t^j\right);$$

therefore σ_x is one-one and so it is a monomorphism. By Exercise 2, page 207, $t = \sigma_x^{-1}(x)$ is an indeterminate over A.

Q.E.D.

As[5] the term suggests, the substitution mapping (or, homomorphism) is closely related to ordinary substitution for variables in the elementary study of polynomial functions. To illustrate this point, we consider

---

[5] The balance of Section 8 should be skipped in a first reading of this chapter.

EXAMPLE 9. As in Example 8, let $f(x) = 2 + 3x - x^2 + \frac{1}{2}x^3$, $x \in Q$. The value of f at 2 is

$$f(2) = 2 + 3 \cdot 2 - 2^2 + \frac{1}{2} \cdot 2^3 = 8.$$

We have seen that f may be regarded as an element in the polynomial domain $A^*[u]$ where $A^* = \mathscr{CF}(Q)$ and the identity function $u : Q \longrightarrow Q$ is an indeterminate over A^*. An easy computation shows that

$$f = c_2 + c_3u - u^2 + c_{1/2}u^3.$$

Now let $\gamma = c_2 \in A^*$. Then

$$\sigma_{c_2}(f) = c_2 + c_3c_2 - c_2^2 + c_{1/2}c_2^3$$
$$= c_2 + c_6 - c_4 + c_4$$
$$= c_8,$$

and for each $x \in Q$, the function $\sigma_{c_2}(f)$ has the value

$$(\sigma_{c_2}(f))(x) = c_8(x) = 8.$$

Thus, the substitution of 2 for x yields the function value 8. On the other hand, the application of the substitution homomorphism σ_{c_2} to the function f yields the constant function c_8 and for each $x \in Q$, $c_8(x) = 8$.

The concept of substitution homomorphism leads directly to another characterization of indeterminates which is valuable in studying *sets* of indeterminates.

THEOREM 11. Let A be a unitary subring of a ring B, and let $t \in B$. The element t is an indeterminate over A if and only if for each ring K, for each $\gamma \in K$, and for each homomorphism $\eta : A \longrightarrow K$, there exists a homomorphism $\tau : A[t] \longrightarrow K$ such that
    (i) $\tau \mid A = \eta$, and
    (ii) $\tau(t) = \gamma$.

Diagrammatically, Theorem 11 is represented by Figure 20 where $\iota : A \longrightarrow A[t]$ is the inclusion mapping and $\tau \circ \iota = \eta$.

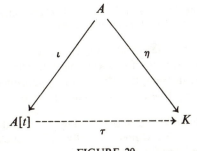

FIGURE 20

We remark that the ring K is subject to no condition whatever. But if K is required to be an extension ring of A and $\eta : A \longrightarrow K$ is the inclusion mapping, then, as the reader should verify, τ is simply the substitution homomorphism σ_γ. Thus by specializing K and η appropriately, Theorem 11 becomes Theorem 10.

    Proof: Suppose t is an indeterminate over A, and $\eta : A \longrightarrow K$ is a homomorphism. Define $\tau : A[t] \longrightarrow K$ by

$$\tau\left(\sum_{i=0}^{m} a_i t^i\right) = \sum_{i=0}^{m} \eta(a_i)\gamma^i.$$

Since t is an indeterminate over A, τ must be a mapping, hence it is a homomorphism (details?). The definition of τ shows that $\tau \,|\, A = \eta$ and that $\tau(t) = \gamma$.

    Conversely, assume that the condition of the theorem holds. Let x be an indeterminate over A, take $K = A[x]$ and let $\eta : A \longrightarrow A[x]$ be the inclusion mapping.

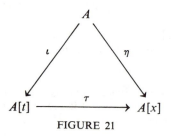

FIGURE 21

By hypothesis there exists a homomorphism $\tau : A[t] \longrightarrow A[x]$ such that $\tau \,|\, A = \eta$ and $\tau(t) = x$. As in the proof of Theorem 10, it is easy to see that τ is a monomorphism. Hence t is an indeterminate over A.

Q.E.D.

    Exercise: Use the first half of Theorem 11 to give an alternative proof that the element t, in the second half, is an indeterminate over A.

As remarked earlier, one virtue of the characterization of indeterminates given by Theorem 11 is that it affords a simple extension to *sets* of indeterminates over a ring. To illustrate, let A be a unitary subring of a (commutative) ring B and let $T = \{x, y\}$ be a two-element subset of B.

    DEFINITION 8. The subring of B generated by $A \cup T$ is the *ring of polynomials in x and y over A* and it is denoted by "$A[x, y]$." Each element in $A[x, y]$ is a *polynomial in x and y with coefficients in A.*

By reasoning similar to that of pages 203 and 204, one sees that each element $p \in A[x, y]$ is a sum

$$p = \sum_{i=0}^{m} \sum_{j=0}^{n} a_{ij} x^i y^j,$$

where m and n are nonnegative integers and $a_{ij} \in A$, $0 \leq i \leq m$, $0 \leq j \leq n$.

DEFINITION 9. $T = \{x, y\}$ is a *set of two indeterminates over* A if and only if for each ring K, for each mapping $\sigma : T \longrightarrow K$ and for each homomorphism $\eta : A \longrightarrow K$, there is a homomorphism $\tau : A[x, y] \longrightarrow K$ such that
(i) $\tau \,|\, A = \eta$, and
(ii) $\tau(x) = \sigma(x)$, $\tau(y) = \sigma(y)$.

If Definition 9 is restated for the case that $T = \{x\}$ is a one-element subset of B, then, aside from trivial differences, it becomes the condition in Theorem 11.

Exercises

1. Show that $A[x]$ and $A[y]$ are unitary subrings of $A[x, y]$ and that $A[x, y] = A[x][y] = A[y][x]$.
2. Prove: If $T = \{x, y\}$ is a set of two indeterminates over A, then x is an indeterminate over $A[y]$ and y is an indeterminate over $A[x]$. Hence, show that each of x and y is an indeterminate over A.

In Exercises 3 and 4, $T = \{x, y\}$ is a set of two indeterminates over A.

3. Show that $p \in A[x, y]$, $p = \sum_{i=0}^{m} \sum_{j=0}^{n} a_{ij} x^i y^j$, is the zero element of $A[x, y]$ if and only if $a_{ij} = 0$, $0 \leq i \leq m$ and $0 \leq j \leq n$.

4. Prove that $p, q \in A[x, y]$ are equal, $p = \sum_{i=0}^{m_1} \sum_{j=0}^{n_1} a_{ij} x^i y^j$, $q = \sum_{i=0}^{m_2} \sum_{j=0}^{n_2} b_{ij} x^i y^j$, if and only if $m_1 = m_2$, $n_1 = n_2$ and $a_{ij} = b_{ij}$, $0 \leq i \leq m_1$ and $0 \leq j \leq n_1$.

Definitions 8 and 9 are extended to arbitrary subsets T of B without difficulty. Again, let A be a unitary subring of a ring B and let T be a subset of B.

DEFINITION 8'. The subring of B generated by $A \cup T$ is the *ring of polynomials in T over A* and it is denoted by "$A[T]$." Each element in $A[T]$ is a *polynomial in T with coefficients in A*.

In case $T = \{x_1, x_2, \ldots, x_k\}$ is finite, then we write "$A[x_1, x_2, \ldots, x_k]$" in place of "$A[T]$" and the elements of $A[x_1, x_2, \ldots, x_k]$ are sums

$$\sum_{i_k=0}^{n_k} \cdots \sum_{i_2=0}^{n_2} \sum_{i_1=0}^{n_1} a_{i_1 i_2 \ldots i_k} x_1^{i_1} x_2^{i_2} \cdots x_k^{i_k}.$$

However, if the cardinality of T is not specified, then $p \in A[T]$ if and only if there exist integers $r, n_1, n_2, \ldots, n_r \in N$, elements $y_1, y_2, \ldots, y_r \in T$ and elements $a_{i_1 i_2 \ldots i_r} \in A$, where $0 \leq i_j \leq n_j$ and $j = 1, 2, \ldots, r$, such that

$$p = \sum_{i_r=0}^{n_r} \cdots \sum_{i_2=0}^{n_2} \sum_{i_1=0}^{n_1} a_{i_1 i_2 \ldots i_r} y_1^{i_1} y_2^{i_2} \cdots y_r^{i_r}.$$

Definition 9 is easily modified into

Definition 9'. T is a *set of indeterminates over A* if and only if for each ring K, for each mapping $\sigma : T \longrightarrow K$ and for each homomorphism $\eta : A \longrightarrow K$, there is a homomorphism $\tau : A[T] \longrightarrow K$ such that

    (i) $\tau \,|\, A = \eta$, and
    (ii) $\tau(x) = \sigma(x)$ for each $x \in T$.

9. *Substitution Homomorphisms*

Let F be a field, and let x be an indeterminate over F. By Theorem 10 it follows that for each extension ring K of F and for each element $\gamma \in K$, the substitution mapping $\sigma_\gamma : F[x] \longrightarrow K$ exists, and in fact, it is a homomorphism. The purpose of the present section is to study substitution homomorphisms in the case that K is an extension *field* of F. The results obtained here will be applied, in Section 10, to the investigation of roots of polynomials in $F[x]$.

Before proceeding to the main theorem observe that by the definition of σ_γ, the image of $F[x]$ under σ_γ is the polynomial domain $F[\gamma]$; i.e.,

$$\sigma_\gamma[F[x]] = F[\gamma].$$

Theorem 12. Let F be a field, let x be an indeterminate over F, and let K be an extension field of F. Further, let σ_γ be the substitution homomorphism determined by $\gamma \in K$. Then

    (i) $\ker (\sigma_\gamma)$ is a prime ideal in $F[x]$, and it is never the unit ideal.
    (ii) if $\ker (\sigma_\gamma) \neq \{0\}$, then
a) $\ker (\sigma_\gamma)$ is a maximal ideal in $F[x]$ and the image $F[\gamma]$ is a subfield of K.
b) $\ker (\sigma_\gamma)$ is a principal ideal generated by one and only one monic, irreducible polynomial.
c) If the monic, irreducible polynomial which generates $\ker (\sigma_\gamma)$ is of degree m, then for each $u \in F[\gamma]$ there exists a unique, ordered m-tuple $(b_0, b_1, \ldots, b_{m-1})$ of elements in F such that

$$u = b_0 + b_1 \gamma + \cdots + b_{m-1} \gamma^{m-1}.$$

Proof: (i) By Theorem 31, Chapter 4, page 180

    (22) $F[\gamma] \cong F[x]/\ker (\sigma_\gamma).$

Since $F[\gamma]$ is an integral domain, Theorem 38, Chapter 4 (page 194) assures us that ker (σ_γ) is a prime ideal.

Since $F[x]$ is an integral domain, the zero ideal in $F[x]$ is prime. Hence one of the possibilities is ker $(\sigma_\gamma) = \{0\}$. In this case $F[\gamma]$ and $F[x]$ are isomorphic, and since $\sigma_\gamma(x) = \gamma$, γ is an indeterminate over F.

On the other hand, ker (σ_γ) can never be the unit ideal. For, since $F \subset F[\gamma]$, $F[\gamma]$ contains at least two elements. Now if ker (σ_γ) were the unit ideal $(= F[x])$, $F[x]/\ker (\sigma_\gamma)$ would be the zero ring, and so $F[\gamma]$ would contain exactly one element. This contradiction shows that ker (σ_γ) is not the unit ideal.

We also deduce that ker (σ_γ) contains no constant, nonzero polynomials. Indeed, if $p = a_0$, $a_0 \neq 0$, and if $p \in \ker (\sigma_\gamma)$, then setting $q = a_0^{-1}$ we have

$$qp = a_0^{-1}a_0 = 1 \in \ker (\sigma_\gamma),$$

whence ker (σ_γ) is the unit ideal, a contradiction. Therefore, ker (σ_γ) contains no constant nonzero polynomials.

(ii) Henceforth assume ker $(\sigma_\gamma) \neq \{0\}$.

a) Since $F[x]$ is a PID, each prime, nonzero, nonunit ideal in $F[x]$ is maximal (Theorem 37, Chapter 4, page 194). Hence ker (σ_γ) is maximal, and therefore by Theorem 38, Chapter 4, and by (22), $F[\gamma]$ is a field. Consequently, $F[\gamma]$ is a subfield of K.

b) Again, from the fact that $F[x]$ is a PID, we know ker (σ_γ) is generated by a polynomial $q \in F[x]$. By (i), $\partial q = m > 0$. Since ker (σ_γ) is a prime ideal, its generator q must be an irreducible polynomial.

If p is an irreducible polynomial in ker (σ_γ), then there is a polynomial $s \in F[x]$ such that $p = sq$. But as p is irreducible, s must be a unit in $F[x]$, i.e., s is a nonzero element in F. Therefore

$$\ker (\sigma_\gamma) = qF[x] = (sq)F[x] = pF[x].$$

We leave it to the reader to complete the proof that there is one and only one monic, irreducible polynomial generating ker (σ_γ) and that this generator is a polynomial of lowest degree in ker (σ_γ).

c) Suppose $g \in F[x]$; throughout the following we shall set

$$g(\gamma) = \sigma_\gamma(g);$$

thus if

$$g = \sum_{i=0}^{n} a_i x^i,$$

then

$$g(\gamma) = \sum_{i=0}^{n} a_i \gamma^i.$$

Now suppose p is the monic, irreducible polynomial generating ker (σ_γ), $\partial p = m$. If $u \in F[\gamma]$, there is a polynomial $f \in F[x]$

such that $\sigma_\gamma(f) = u$. By the division algorithm in $F[x]$, there exist unique polynomials q and r in $F[x]$ such that

$$f = pq + r,$$

where

$$r = \sum_{i=0}^{m-1} b_i x^i.$$

Then

$$\begin{aligned}
u &= \sigma_\gamma(f) \\
&= \sigma_\gamma(pq + r) \\
&= \sigma_\gamma(p)\sigma_\gamma(q) + \sigma_\gamma(r) \\
&= p(\gamma)q(\gamma) + r(\gamma).
\end{aligned}$$

But since $p \in \ker (\sigma_\gamma)$, $p(\gamma) = \sigma_\gamma(p) = 0$, and therefore

$$u = r(\gamma) = \sum_{i=0}^{m-1} b_i \gamma^i.$$

This establishes the first part of c).

To prove uniqueness, suppose

$$\sum_{i=0}^{m-1} b_i \gamma^i = \sum_{i=0}^{m-1} c_i \gamma^i;$$

setting $d_i = b_i - c_i$, $0 \leq i \leq m - 1$, and

$$z = \sum_{i=0}^{m-1} d_i x^i,$$

we find that

$$\sigma_\gamma(z) = \sum_{i=0}^{m-1} d_i \gamma^i = 0,$$

hence $z \in \ker (\sigma_\gamma)$. But since p is a polynomial of lowest degree in $\ker (\sigma_\gamma)$, it follows that $z = 0$. Therefore $d_i = 0$ and so $b_i = c_i$, $0 \leq i \leq m - 1$.

Q.E.D.

COROLLARY 1.   The monic, irreducible polynomial which generates $\ker (\sigma_\gamma)$ is a polynomial of lowest degree in $\ker (\sigma_\gamma)$.

Proof: Part (ii) b) of Theorem 12.

Q.E.D.

COROLLARY 2.   If $\gamma \in F$, then $\ker (\sigma_\gamma) = (x - \gamma)F[x]$. Hence $p \in \ker (\sigma_\gamma)$ if and only if there is a $q \in F[x]$ such that $p = (x - \gamma)q$.

Proof: If $\gamma \in F$, then $\sigma_\gamma(x - \gamma) = \gamma - \gamma = 0$, hence the monic, irreducible polynomial $x - \gamma$ is an element in ker (σ_γ). Since $x - \gamma$ is evidently a polynomial of lowest degree in ker (σ_γ) Corollary 1 and Theorem 12 imply that $x - \gamma$ generates ker (σ_γ).

The second statement is left to the reader.

Q.E.D.

EXAMPLE 10.   Let $F = Q$, the field of rational numbers, let C be the field of complex numbers and let $\gamma = i$ where $i^2 = -1$. The polynomial $1 + x^2 \in Q[x]$ is irreducible (why?) and

$$\sigma_i(1 + x^2) = 1 + i^2 = 0,$$

so that $1 + x^2$ is a monic, irreducible polynomial in ker (σ_i). Hence

$$\text{ker } (\sigma_i) = (1 + x^2)Q[x].$$

Since ker $(\sigma_i) \neq \{0\}$, it is a maximal ideal in $Q[x]$ and therefore (Theorem 31, Chapter 4, page 180)

$$\sigma_i[Q[x]] = Q[i] \cong Q[x]/\text{ker } (\sigma_i)$$

is a field where

$$Q[i] = \{a + bi \,|\, a, b \in Q\}.$$

Exercises

1. Let $F = Q$, the field of rationals, let $K = R$, the field of reals, and let $\gamma = \sqrt{2}$. Determine ker $(\sigma_{\sqrt{2}})$ and describe $\sigma_{\sqrt{2}}[Q[x]] = Q[\sqrt{2}]$.

2. Let $F = Z_2$ and let K be an extension field of Z_2 containing an element α such that $\bar{1} + \alpha + \alpha^2 = \bar{0}$. Determine ker (σ_α) and describe $Z_2[\alpha]$. (*Note:* We have given no proof of the existence of a field K satisfying the conditions of the exercise. For a discussion of the general topic to which this problem is related, see I. N. Herstein, *Topics in Algebra*, Chapter 5, (Blaisdell, 1964).

10. *Roots of Polynomials*

Throughout this section, F is to be a field. Let $p \in F[x]$ and let $\alpha \in F$. Since $F[x]$ is a UFD, and since $x - \alpha$ is an irreducible polynomial in $F[x]$ (Exercise 7, page 215), there is an integer $k \geq 0$ and a polynomial $q \in F[x]$ such that

$$p = (x - \alpha)^k q,$$

where $x - \alpha$ is not a factor of q.

DEFINITION 10.   An element $\alpha \in F$ is a *root* of $p \in F[x]$ of *multiplicity k*, $k \geq 0$, if and only if

(i) $p = (x - \alpha)^k q$ where $q \in F[x]$, and
(ii) $x - \alpha$ is not a factor of q.

A root of multiplicity one is a *simple* root; a root of multiplicity two is a *double* root; a root of multiplicity three is a *triple* root, etc. The term "root" without designation of multiplicity denotes a root whose multiplicity exceeds zero. If α is a root of multiplicity zero, it is not a root of p.

For example, the polynomial

$$(x - 2)(x^2 - 2)^2(x - 3)^3 \in R[x]$$

has 2 as a simple root, both $\sqrt{2}$ and $-\sqrt{2}$ as double roots, and 3 as a triple root. All real numbers different from 2, $\sqrt{2}$, $-\sqrt{2}$ and 3 are roots of multiplicity zero, i.e., they are not roots of the given polynomial.

THEOREM 13.   Let $p \in F[x]$. An element $\alpha \in F$ is a root of p if and only if $p(\alpha) = \sigma_\alpha(p) = 0$, hence if and only if $p \in \ker(\sigma_\alpha)$.

Proof: If α is a root of p, then by Definition 10, $p = (x - \alpha)^k q$ where $k \geq 1$ and $q \in F[x]$. Hence

$$\begin{aligned} p(\alpha) = \sigma_\alpha(p) &= \sigma_\alpha((x - \alpha)^k q) \\ &= (\sigma_\alpha(x - \alpha))^k \sigma_\alpha(q) \\ &= 0, \end{aligned}$$

since $\sigma_\alpha(x - \alpha) = \alpha - \alpha = 0$. Therefore $p \in \ker(\sigma_\alpha)$.

Conversely, suppose $p \in \ker(\sigma_\alpha)$. By Corollary 2 to Theorem 12, $\ker(\sigma_\alpha) = (x - \alpha)F[x]$, and therefore

$$p = (x - \alpha)r,$$

where $r \in F[x]$. Hence $p = (x - \alpha)^k q$ where $k \geq 1$, $q \in F[x]$ and $x - \alpha$ is not a factor of q. By Definition 10, α is a root of p. Q.E.D.

An important result concerning multiplicities of roots of a polynomial is

THEOREM 14.   Let $p \in F[x]$, $\partial p = m$. Then the sum of the multiplicities of the roots of p, in F, does not exceed m.

Proof: Since $F[x]$ is a UFD, there exist irreducible polynomials $p_i \in F[x]$, and positive integers m_i, $1 \leq i \leq s$, such that

$$p = a p_1^{m_1} p_2^{m_2} \cdots p_s^{m_s},$$

where $a \neq 0$ is an element in F. Further, since the nonzero elements in F are the units in $F[x]$, we may assume that each p_i is monic (why?). If $\partial p_i = n_i$, $1 \leq i \leq s$, then since p_i is irreducible, the corresponding n_i is positive and

$$\begin{aligned} (23) \qquad m = \partial p &= \partial(p_1^{m_1}) + \partial(p_2^{m_2}) + \cdots + \partial(p_s^{m_s}) \\ &= m_1 \partial p_1 + m_2 \partial p_2 + \cdots + m_s \partial p_s \\ &= m_1 n_1 + m_2 n_2 + \cdots + m_s n_s. \end{aligned}$$

Now suppose that $n_1 = n_2 = \cdots = n_r = 1$ where $0 \leq r \leq s$, and that $n_j > 1$, $r + 1 \leq j \leq s$. Then

$$p_i = x - \alpha_i, \quad 1 \leq i \leq r,$$

where $\alpha_i \in F$, and the α_i are mutually distinct. Thus each α_i, $1 \leq i \leq r$, is a root of p of multiplicity m_i and equation (23) shows that

$$m_1 + m_2 + \cdots + m_r \leq m.$$

The equality occurs only in case that p has no irreducible factors of degree greater than 1.

Finally, p has no root in F other than $\alpha_1, \alpha_2, \ldots, \alpha_r$. Indeed, suppose β is a root of p where $\beta \in F$. Then $x - \beta$ is a factor of p and since $F[x]$ is a UFD, $x - \beta$ is a factor of one of the irreducible polynomials $p_1, p_2, \ldots, p_s$. Clearly, $x - \beta$ is not a factor of any of the polynomials $p_{r+1}, \ldots, p_s$. Hence $x - \beta$ is a factor of one of the $p_j = x - \alpha_j$, $1 \leq j \leq r$. Since $x - \beta$ and $x - \alpha_j$ are both monic, irreducible and of first degree, $x - \beta = x - \alpha_j$, whence $\beta = \alpha_j$.

It now follows that the sum of the multiplicities of the roots of p, in F, is $m_1 + m_2 + \cdots + m_r$ and the theorem is proved.

Q.E.D.

Consider the polynomial $x^2 - 2 \in Q[x]$. Since there exists no $\alpha \in R$ (= reals) such that

$$x^2 - 2 = (x - \alpha)^k q,$$

where k is an integer, $k \geq 1$, and $q \in Q[x]$, we know that $x^2 - 2$ has no root in Q. On the other hand, there *does* exist an element, namely $\sqrt{2}$, in the field R such that

$$x^2 - 2 = (x - \sqrt{2})q,$$

where $q \in R[x]$. This example—and others like it—suggests that the concept of root be generalized so that elements in extension fields of the coefficient field be regarded as candidates for roothood.

Let K be an extension field of a field F, and assume x is an indeterminate over K. Then (Exercise 4, page 206) x is also an indeterminate over F.

DEFINITION 11.   An element $\alpha \in K$ is a *root* of $p \in F[x]$ of *multiplicity m*, $m \geq 0$, if and only if there is a polynomial $q \in K[x]$ such that
(i) $p = (x - \alpha)^m q$, and
(ii) $x - \alpha$ is not a factor of q in $K[x]$.

The reader should compare Definition 11, carefully, with Definition 10. The terminology of *simple*, *double* and *triple* roots is carried over without modification. For instance, since the polynomial

$$p = (x^2 + 1)^2(x - 2) \in Q[x]$$

has the factorization
$$(x - i)^2(x + i)^2(x - 2) = (x - i)^2 q,$$
where $q = (x + i)^2(x - 2) \in C[x]$, C being the field of complex numbers, and since $x - i$ is not a factor of q in $C[x]$, it follows that i is a double root op p in C.

Theorems 13 and 14 are now generalized as follows:

THEOREM 15. Let $p \in F[x]$ and let K be an extension field of F. An element $\alpha \in K$ is a root of p if and only if $p \in \ker(\sigma_\alpha)$ where $\sigma_\alpha : F[x] \longrightarrow K$ is the substitution homomorphism.

THEOREM 16. Let $p \in F[x]$, $\partial p = n$, and let K be an extension field of F. Then the sum of the multiplicities of the roots of p, in K, does not exceed n.

In proving these new theorems, one need only observe that since x is an indeterminate over K, hence over F, $p \in F[x]$ implies $p \in K[x]$. The proofs of Theorems 13 and 14 now carry over, word for word, the sole alteration being the replacement of F by K. The reader should go through the details.

Exercises

1. Prove Theorems 15 and 16.

2. Let K be an extension field of F, and let $p = \sum\limits_{i=0}^{m} a_i x^i \in F[x]$, $a_m \neq 0$. Let $p' = \sum\limits_{i=0}^{m} b_i x^i$ where $b_i = a_i a_m^{-1}$, $0 \leq i \leq m$. Prove that $\alpha \in K$ is a root of p if and only if α is a root of p'. (*Note:* p' is a monic polynomial.)

3. Let K be an extension field of F, and let $p = qr$ where $p, q, r \in F[x]$. Prove: An element $\alpha \in K$ is a root of p if and only if α is a root of q or of r. Hence show that α is a root of p if and only if α is a root of an irreducible factor of p.

4. Prove that the complex roots of a polynomial over R occur in conjugate pairs.

Index

241

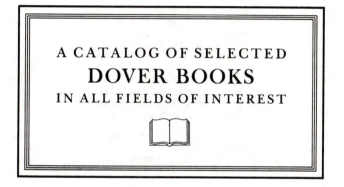

A CATALOG OF SELECTED
DOVER BOOKS
IN ALL FIELDS OF INTEREST

A CATALOG OF SELECTED DOVER
BOOKS IN ALL FIELDS OF INTEREST

DRAWINGS OF REMBRANDT, edited by Seymour Slive. Updated Lippmann, Hofstede de Groot edition, with definitive scholarly apparatus. All portraits, biblical sketches, landscapes, nudes. Oriental figures, classical studies, together with selection of work by followers. 550 illustrations. Total of 630pp. 9⅛ × 12¼.
21485-0, 21486-9 Pa., Two-vol. set $25.00

GHOST AND HORROR STORIES OF AMBROSE BIERCE, Ambrose Bierce. 24 tales vividly imagined, strangely prophetic, and decades ahead of their time in technical skill: "The Damned Thing," "An Inhabitant of Carcosa," "The Eyes of the Panther," "Moxon's Master," and 20 more. 199pp. 5⅜ × 8½. 20767-6 Pa. $3.95

ETHICAL WRITINGS OF MAIMONIDES, Maimonides. Most significant ethical works of great medieval sage, newly translated for utmost precision, readability. Laws Concerning Character Traits, Eight Chapters, more. 192pp. 5⅜ × 8½.
24522-5 Pa. $4.50

THE EXPLORATION OF THE COLORADO RIVER AND ITS CANYONS, J. W. Powell. Full text of Powell's 1,000-mile expedition down the fabled Colorado in 1869. Superb account of terrain, geology, vegetation, Indians, famine, mutiny, treacherous rapids, mighty canyons, during exploration of last unknown part of continental U.S. 400pp. 5⅜ × 8½. 20094-9 Pa. $6.95

HISTORY OF PHILOSOPHY, Julián Marías. Clearest one-volume history on the market. Every major philosopher and dozens of others, to Existentialism and later. 505pp. 5⅜ × 8½. 21739-6 Pa. $8.50

ALL ABOUT LIGHTNING, Martin A. Uman. Highly readable non-technical survey of nature and causes of lightning, thunderstorms, ball lightning, St. Elmo's Fire, much more. Illustrated. 192pp. 5⅜ × 8½. 25237-X Pa. $5.95

SAILING ALONE AROUND THE WORLD, Captain Joshua Slocum. First man to sail around the world, alone, in small boat. One of great feats of seamanship told in delightful manner. 67 illustrations. 294pp. 5⅜ × 8½. 20326-3 Pa. $4.95

LETTERS AND NOTES ON THE MANNERS, CUSTOMS AND CONDITIONS OF THE NORTH AMERICAN INDIANS, George Catlin. Classic account of life among Plains Indians: ceremonies, hunt, warfare, etc. 312 plates. 572pp. of text. 6⅛ × 9¼. 22118-0, 22119-9 Pa. Two-vol. set $15.90

ALASKA: The Harriman Expedition, 1899, John Burroughs, John Muir, et al. Informative, engrossing accounts of two-month, 9,000-mile expedition. Native peoples, wildlife, forests, geography, salmon industry, glaciers, more. Profusely illustrated. 240 black-and-white line drawings. 124 black-and-white photographs. 3 maps. Index. 576pp. 5⅜ × 8½. 25109-8 Pa. $11.95

CATALOG OF DOVER BOOKS

THE BOOK OF BEASTS: Being a Translation from a Latin Bestiary of the Twelfth Century, T. H. White. Wonderful catalog real and fanciful beasts: manticore, griffin, phoenix, amphivius, jaculus, many more. White's witty erudite commentary on scientific, historical aspects. Fascinating glimpse of medieval mind. Illustrated. 296pp. 5⅜ × 8¼. (Available in U.S. only) 24609-4 Pa. $5.95

FRANK LLOYD WRIGHT: ARCHITECTURE AND NATURE With 160 Illustrations, Donald Hoffmann. Profusely illustrated study of influence of nature—especially prairie—on Wright's designs for Fallingwater, Robie House, Guggenheim Museum, other masterpieces. 96pp. 9¼ × 10¾. 25098-9 Pa. $7.95

FRANK LLOYD WRIGHT'S FALLINGWATER, Donald Hoffmann. Wright's famous waterfall house: planning and construction of organic idea. History of site, owners, Wright's personal involvement. Photographs of various stages of building. Preface by Edgar Kaufmann, Jr. 100 illustrations. 112pp. 9¼ × 10. 23671-4 Pa. $7.95

YEARS WITH FRANK LLOYD WRIGHT: Apprentice to Genius, Edgar Tafel. Insightful memoir by a former apprentice presents a revealing portrait of Wright the man, the inspired teacher, the greatest American architect. 372 black-and-white illustrations. Preface. Index. vi + 228pp. 8¼ × 11. 24801-1 Pa. $9.95

THE STORY OF KING ARTHUR AND HIS KNIGHTS, Howard Pyle. Enchanting version of King Arthur fable has delighted generations with imaginative narratives of exciting adventures and unforgettable illustrations by the author. 41 illustrations. xviii + 313pp. 6⅛ × 9¼. 21445-1 Pa. $6.50

THE GODS OF THE EGYPTIANS, E. A. Wallis Budge. Thorough coverage of numerous gods of ancient Egypt by foremost Egyptologist. Information on evolution of cults, rites and gods; the cult of Osiris; the Book of the Dead and its rites; the sacred animals and birds; Heaven and Hell; and more. 956pp. 6⅛ × 9¼. 22055-9, 22056-7 Pa., Two-vol. set $20.00

A THEOLOGICO-POLITICAL TREATISE, Benedict Spinoza. Also contains unfinished Political Treatise. Great classic on religious liberty, theory of government on common consent. R. Elwes translation. Total of 421pp. 5⅜ × 8½. 20249-6 Pa. $6.95

INCIDENTS OF TRAVEL IN CENTRAL AMERICA, CHIAPAS, AND YUCATAN, John L. Stephens. Almost single-handed discovery of Maya culture; exploration of ruined cities, monuments, temples; customs of Indians. 115 drawings. 892pp. 5⅜ × 8½. 22404-X, 22405-8 Pa., Two-vol. set $15.90

LOS CAPRICHOS, Francisco Goya. 80 plates of wild, grotesque monsters and caricatures. Prado manuscript included. 183pp. 6⅛ × 9⅜. 22384-1 Pa. $4.95

AUTOBIOGRAPHY: The Story of My Experiments with Truth, Mohandas K. Gandhi. Not hagiography, but Gandhi in his own words. Boyhood, legal studies, purification, the growth of the Satyagraha (nonviolent protest) movement. Critical, inspiring work of the man who freed India. 480pp. 5⅜ × 8½. (Available in U.S. only) 24593-4 Pa. $6.95

CATALOG OF DOVER BOOKS

ILLUSTRATED DICTIONARY OF HISTORIC ARCHITECTURE, edited by Cyril M. Harris. Extraordinary compendium of clear, concise definitions for over 5,000 important architectural terms complemented by over 2,000 line drawings. Covers full spectrum of architecture from ancient ruins to 20th-century Modernism. Preface. 592pp. 7½ × 9⅝. 24444-X Pa. $14.95

THE NIGHT BEFORE CHRISTMAS, Clement Moore. Full text, and woodcuts from original 1848 book. Also critical, historical material. 19 illustrations. 40pp. 4⅝ × 6. 22797-9 Pa. $2.25

THE LESSON OF JAPANESE ARCHITECTURE: 165 Photographs, Jiro Harada. Memorable gallery of 165 photographs taken in the 1930's of exquisite Japanese homes of the well-to-do and historic buildings. 13 line diagrams. 192pp. 8⅜ × 11¼. 24778-3 Pa. $8.95

THE AUTOBIOGRAPHY OF CHARLES DARWIN AND SELECTED LETTERS, edited by Francis Darwin. The fascinating life of eccentric genius composed of an intimate memoir by Darwin (intended for his children); commentary by his son, Francis; hundreds of fragments from notebooks, journals, papers; and letters to and from Lyell, Hooker, Huxley, Wallace and Henslow. xi + 365pp. 5⅝ × 8.
20479-0 Pa. $6.95

WONDERS OF THE SKY: Observing Rainbows, Comets, Eclipses, the Stars and Other Phenomena, Fred Schaaf. Charming, easy-to-read poetic guide to all manner of celestial events visible to the naked eye. Mock suns, glories, Belt of Venus, more. Illustrated. 299pp. 5¼ × 8¼. 24402-4 Pa. $7.95

BURNHAM'S CELESTIAL HANDBOOK, Robert Burnham, Jr. Thorough guide to the stars beyond our solar system. Exhaustive treatment. Alphabetical by constellation: Andromeda to Cetus in Vol. 1; Chamaeleon to Orion in Vol. 2; and Pavo to Vulpecula in Vol. 3. Hundreds of illustrations. Index in Vol. 3. 2,000pp. 6⅛ × 9¼. 23567-X, 23568-8, 23673-0 Pa., Three-vol. set $38.85

STAR NAMES: Their Lore and Meaning, Richard Hinckley Allen. Fascinating history of names various cultures have given to constellations and literary and folkloristic uses that have been made of stars. Indexes to subjects. Arabic and Greek names. Biblical references. Bibliography. 563pp. 5⅜ × 8½. 21079-0 Pa. $7.95

THIRTY YEARS THAT SHOOK PHYSICS: The Story of Quantum Theory, George Gamow. Lucid, accessible introduction to influential theory of energy and matter. Careful explanations of Dirac's anti-particles, Bohr's model of the atom, much more. 12 plates. Numerous drawings. 240pp. 5⅜ × 8½. 24895-X Pa. $4.95

CHINESE DOMESTIC FURNITURE IN PHOTOGRAPHS AND MEASURED DRAWINGS, Gustav Ecke. A rare volume, now affordably priced for antique collectors, furniture buffs and art historians. Detailed review of styles ranging from early Shang to late Ming. Unabridged republication. 161 black-and-white drawings, photos. Total of 224pp. 8⅜ × 11¼. (Available in U.S. only) 25171-3 Pa. $12.95

VINCENT VAN GOGH: A Biography, Julius Meier-Graefe. Dynamic, penetrating study of artist's life, relationship with brother, Theo, painting techniques, travels, more. Readable, engrossing. 160pp. 5⅜ × 8½. (Available in U.S. only)
25253-1 Pa. $3.95

HOW TO WRITE, Gertrude Stein. Gertrude Stein claimed anyone could understand her unconventional writing—here are clues to help. Fascinating improvisations, language experiments, explanations illuminate Stein's craft and the art of writing. Total of 414pp. 4⅝ × 6⅜. 23144-5 Pa. $5.95

ADVENTURES AT SEA IN THE GREAT AGE OF SAIL: Five Firsthand Narratives, edited by Elliot Snow. Rare true accounts of exploration, whaling, shipwreck, fierce natives, trade, shipboard life, more. 33 illustrations. Introduction. 353pp. 5⅜ × 8½. 25177-2 Pa. $7.95

THE HERBAL OR GENERAL HISTORY OF PLANTS, John Gerard. Classic descriptions of about 2,850 plants—with over 2,700 illustrations—includes Latin and English names, physical descriptions, varieties, time and place of growth, more. 2,706 illustrations. xlv + 1,678pp. 8½ × 12¼. 23147-X Cloth. $75.00

DOROTHY AND THE WIZARD IN OZ, L. Frank Baum. Dorothy and the Wizard visit the center of the Earth, where people are vegetables, glass houses grow and Oz characters reappear. Classic sequel to *Wizard of Oz*. 256pp. 5⅝ × 8. 24714-7 Pa. $4.95

SONGS OF EXPERIENCE: Facsimile Reproduction with 26 Plates in Full Color, William Blake. This facsimile of Blake's original "Illuminated Book" reproduces 26 full-color plates from a rare 1826 edition. Includes "The Tyger," "London," "Holy Thursday," and other immortal poems. 26 color plates. Printed text of poems. 48pp. 5¼ × 7. 24636-1 Pa. $3.50

SONGS OF INNOCENCE, William Blake. The first and most popular of Blake's famous "Illuminated Books," in a facsimile edition reproducing all 31 brightly colored plates. Additional printed text of each poem. 64pp. 5¼ × 7. 22764-2 Pa. $3.50

PRECIOUS STONES, Max Bauer. Classic, thorough study of diamonds, rubies, emeralds, garnets, etc.: physical character, occurrence, properties, use, similar topics. 20 plates, 8 in color. 94 figures. 659pp. 6⅛ × 9¼. 21910-0, 21911-9 Pa., Two-vol. set $15.90

ENCYCLOPEDIA OF VICTORIAN NEEDLEWORK, S. F. A. Caulfeild and Blanche Saward. Full, precise descriptions of stitches, techniques for dozens of needlecrafts—most exhaustive reference of its kind. Over 800 figures. Total of 679pp. 8⅛ × 11. Two volumes. Vol. 1 22800-2 Pa. $11.95
Vol. 2 22801-0 Pa. $11.95

THE MARVELOUS LAND OF OZ, L. Frank Baum. Second Oz book, the Scarecrow and Tin Woodman are back with hero named Tip, Oz magic. 136 illustrations. 287pp. 5⅝ × 8½. 20692-0 Pa. $5.95

WILD FOWL DECOYS, Joel Barber. Basic book on the subject, by foremost authority and collector. Reveals history of decoy making and rigging, place in American culture, different kinds of decoys, how to make them, and how to use them. 140 plates. 156pp. 7⅞ × 10¾. 20011-6 Pa. $8.95

HISTORY OF LACE, Mrs. Bury Palliser. Definitive, profusely illustrated chronicle of lace from earliest times to late 19th century. Laces of Italy, Greece, England, France, Belgium, etc. Landmark of needlework scholarship. 266 illustrations. 672pp. 6⅛ × 9¼. 24742-2 Pa. $14.95

ILLUSTRATED GUIDE TO SHAKER FURNITURE, Robert Meader. All furniture and appurtenances, with much on unknown local styles. 235 photos. 146pp. 9 × 12.   22819-3 Pa. $7.95

WHALE SHIPS AND WHALING: A Pictorial Survey, George Francis Dow. Over 200 vintage engravings, drawings, photographs of barks, brigs, cutters, other vessels. Also harpoons, lances, whaling guns, many other artifacts. Comprehensive text by foremost authority. 207 black-and-white illustrations. 288pp. 6 × 9.
24808-9 Pa. $8.95

THE BERTRAMS, Anthony Trollope. Powerful portrayal of blind self-will and thwarted ambition includes one of Trollope's most heartrending love stories. 497pp. 5⅜ × 8½.   25119-5 Pa. $8.95

ADVENTURES WITH A HAND LENS, Richard Headstrom. Clearly written guide to observing and studying flowers and grasses, fish scales, moth and insect wings, egg cases, buds, feathers, seeds, leaf scars, moss, molds, ferns, common crystals, etc.—all with an ordinary, inexpensive magnifying glass. 209 exact line drawings aid in your discoveries. 220pp. 5⅜ × 8½.   23330-8 Pa. $3.95

RODIN ON ART AND ARTISTS, Auguste Rodin. Great sculptor's candid, wide-ranging comments on meaning of art; great artists; relation of sculpture to poetry, painting, music; philosophy of life, more. 76 superb black-and-white illustrations of Rodin's sculpture, drawings and prints. 119pp. 8⅜ × 11¼.   24487-3 Pa. $6.95

FIFTY CLASSIC FRENCH FILMS, 1912–1982: A Pictorial Record, Anthony Slide. Memorable stills from Grand Illusion, Beauty and the Beast, Hiroshima, Mon Amour, many more. Credits, plot synopses, reviews, etc. 160pp. 8¼ × 11.
25256-6 Pa. $11.95

THE PRINCIPLES OF PSYCHOLOGY, William James. Famous long course complete, unabridged. Stream of thought, time perception, memory, experimental methods; great work decades ahead of its time. 94 figures. 1,391pp. 5⅜ × 8½.
20381-6, 20382-4 Pa., Two-vol. set $19.90

BODIES IN A BOOKSHOP, R. T. Campbell. Challenging mystery of blackmail and murder with ingenious plot and superbly drawn characters. In the best tradition of British suspense fiction. 192pp. 5⅜ × 8½.   24720-1 Pa. $3.95

CALLAS: PORTRAIT OF A PRIMA DONNA, George Jellinek. Renowned commentator on the musical scene chronicles incredible career and life of the most controversial, fascinating, influential operatic personality of our time. 64 black-and-white photographs. 416pp. 5⅜ × 8¼.   25047-4 Pa. $7.95

GEOMETRY, RELATIVITY AND THE FOURTH DIMENSION, Rudolph Rucker. Exposition of fourth dimension, concepts of relativity as Flatland characters continue adventures. Popular, easily followed yet accurate, profound. 141 illustrations. 133pp. 5⅜ × 8½.   23400-2 Pa. $3.95

HOUSEHOLD STORIES BY THE BROTHERS GRIMM, with pictures by Walter Crane. 53 classic stories—Rumpelstiltskin, Rapunzel, Hansel and Gretel, the Fisherman and his Wife, Snow White, Tom Thumb, Sleeping Beauty, Cinderella, and so much more—lavishly illustrated with original 19th century drawings. 114 illustrations. x + 269pp. 5⅜ × 8½.   21080-4 Pa. $4.50

SUNDIALS, Albert Waugh. Far and away the best, most thorough coverage of ideas, mathematics concerned, types, construction, adjusting anywhere. Over 100 illustrations. 230pp. 5⅜ × 8½. 22947-5 Pa. $4.50

PICTURE HISTORY OF THE NORMANDIE: With 190 Illustrations, Frank O. Braynard. Full story of legendary French ocean liner: Art Deco interiors, design innovations, furnishings, celebrities, maiden voyage, tragic fire, much more. Extensive text. 144pp. 8⅞ × 11¼. 25257-4 Pa. $9.95

THE FIRST AMERICAN COOKBOOK: A Facsimile of "American Cookery," 1796, Amelia Simmons. Facsimile of the first American-written cookbook published in the United States contains authentic recipes for colonial favorites— pumpkin pudding, winter squash pudding, spruce beer, Indian slapjacks, and more. Introductory Essay and Glossary of colonial cooking terms. 80pp. 5⅜ × 8½. 24710-4 Pa. $3.50

101 PUZZLES IN THOUGHT AND LOGIC, C. R. Wylie, Jr. Solve murders and robberies, find out which fishermen are liars, how a blind man could possibly identify a color—purely by your own reasoning! 107pp. 5⅜ × 8½. 20367-0 Pa. $2.50

THE BOOK OF WORLD-FAMOUS MUSIC—CLASSICAL, POPULAR AND FOLK, James J. Fuld. Revised and enlarged republication of landmark work in musico-bibliography. Full information about nearly 1,000 songs and compositions including first lines of music and lyrics. New supplement. Index. 800pp. 5⅜ × 8¼. 24857-7 Pa. $14.95

ANTHROPOLOGY AND MODERN LIFE, Franz Boas. Great anthropologist's classic treatise on race and culture. Introduction by Ruth Bunzel. Only inexpensive paperback edition. 255pp. 5⅜ × 8½. 25245-0 Pa. $5.95

THE TALE OF PETER RABBIT, Beatrix Potter. The inimitable Peter's terrifying adventure in Mr. McGregor's garden, with all 27 wonderful, full-color Potter illustrations. 55pp. 4¼ × 5½. (Available in U.S. only) 22827-4 Pa. $1.75

THREE PROPHETIC SCIENCE FICTION NOVELS, H. G. Wells. *When the Sleeper Wakes, A Story of the Days to Come* and *The Time Machine* (full version). 335pp. 5⅜ × 8½. (Available in U.S. only) 20605-X Pa. $5.95

APICIUS COOKERY AND DINING IN IMPERIAL ROME, edited and translated by Joseph Dommers Vehling. Oldest known cookbook in existence offers readers a clear picture of what foods Romans ate, how they prepared them, etc. 49 illustrations. 301pp. 6⅛ × 9¼. 23563-7 Pa. $6.50

SHAKESPEARE LEXICON AND QUOTATION DICTIONARY, Alexander Schmidt. Full definitions, locations, shades of meaning of every word in plays and poems. More than 50,000 exact quotations. 1,485pp. 6½ × 9¼. 22726-X, 22727-8 Pa., Two-vol. set $27.90

THE WORLD'S GREAT SPEECHES, edited by Lewis Copeland and Lawrence W. Lamm. Vast collection of 278 speeches from Greeks to 1970. Powerful and effective models; unique look at history. 842pp. 5⅜ × 8½. 20468-5 Pa. $11.95

THE BLUE FAIRY BOOK, Andrew Lang. The first, most famous collection, with many familiar tales: Little Red Riding Hood, Aladdin and the Wonderful Lamp, Puss in Boots, Sleeping Beauty, Hansel and Gretel, Rumpelstiltskin; 37 in all. 138 illustrations. 390pp. 5⅜ × 8½. 21437-0 Pa. $5.95

THE STORY OF THE CHAMPIONS OF THE ROUND TABLE, Howard Pyle. Sir Launcelot, Sir Tristram and Sir Percival in spirited adventures of love and triumph retold in Pyle's inimitable style. 50 drawings, 31 full-page. xviii + 329pp. 6½ × 9¼. 21883-X Pa. $6.95

AUDUBON AND HIS JOURNALS, Maria Audubon. Unmatched two-volume portrait of the great artist, naturalist and author contains his journals, an excellent biography by his granddaughter, expert annotations by the noted ornithologist, Dr. Elliott Coues, and 37 superb illustrations. Total of 1,200pp. 5⅜ × 8.
Vol. I 25143-8 Pa. $8.95
Vol. II 25144-6 Pa. $8.95

GREAT DINOSAUR HUNTERS AND THEIR DISCOVERIES, Edwin H. Colbert. Fascinating, lavishly illustrated chronicle of dinosaur research, 1820's to 1960. Achievements of Cope, Marsh, Brown, Buckland, Mantell, Huxley, many others. 384pp. 5¼ × 8¼. 24701-5 Pa. $6.95

THE TASTEMAKERS, Russell Lynes. Informal, illustrated social history of American taste 1850's–1950's. First popularized categories Highbrow, Lowbrow, Middlebrow. 129 illustrations. New (1979) afterword. 384pp. 6 × 9.
23993-4 Pa. $6.95

DOUBLE CROSS PURPOSES, Ronald A. Knox. A treasure hunt in the Scottish Highlands, an old map, unidentified corpse, surprise discoveries keep reader guessing in this cleverly intricate tale of financial skullduggery. 2 black-and-white maps. 320pp. 5⅜ × 8½. (Available in U.S. only) 25032-6 Pa. $5.95

AUTHENTIC VICTORIAN DECORATION AND ORNAMENTATION IN FULL COLOR: 46 Plates from "Studies in Design," Christopher Dresser. Superb full-color lithographs reproduced from rare original portfolio of a major Victorian designer. 48pp. 9¼ × 12¼. 25083-0 Pa. $7.95

PRIMITIVE ART, Franz Boas. Remains the best text ever prepared on subject, thoroughly discussing Indian, African, Asian, Australian, and, especially, Northern American primitive art. Over 950 illustrations show ceramics, masks, totem poles, weapons, textiles, paintings, much more. 376pp. 5⅜ × 8. 20025-6 Pa. $6.95

SIDELIGHTS ON RELATIVITY, Albert Einstein. Unabridged republication of two lectures delivered by the great physicist in 1920–21. *Ether and Relativity* and *Geometry and Experience.* Elegant ideas in non-mathematical form, accessible to intelligent layman. vi + 56pp. 5⅜ × 8½. 24511-X Pa. $2.95

THE WIT AND HUMOR OF OSCAR WILDE, edited by Alvin Redman. More than 1,000 ripostes, paradoxes, wisecracks: Work is the curse of the drinking classes, I can resist everything except temptation, etc. 258pp. 5⅜ × 8½. 20602-5 Pa. $4.50

ADVENTURES WITH A MICROSCOPE, Richard Headstrom. 59 adventures with clothing fibers, protozoa, ferns and lichens, roots and leaves, much more. 142 illustrations. 232pp. 5⅜ × 8½. 23471-1 Pa. $3.95

CATALOG OF DOVER BOOKS

PLANTS OF THE BIBLE, Harold N. Moldenke and Alma L. Moldenke. Standard reference to all 230 plants mentioned in Scriptures. Latin name, biblical reference, uses, modern identity, much more. Unsurpassed encyclopedic resource for scholars, botanists, nature lovers, students of Bible. Bibliography. Indexes. 123 black-and-white illustrations. 384pp. 6 × 9. 25069-5 Pa. $8.95

FAMOUS AMERICAN WOMEN: A Biographical Dictionary from Colonial Times to the Present, Robert McHenry, ed. From Pocahontas to Rosa Parks, 1,035 distinguished American women documented in separate biographical entries. Accurate, up-to-date data, numerous categories, spans 400 years. Indices. 493pp. 6½ × 9¼. 24523-3 Pa. $9.95

THE FABULOUS INTERIORS OF THE GREAT OCEAN LINERS IN HISTORIC PHOTOGRAPHS, William H. Miller, Jr. Some 200 superb photographs capture exquisite interiors of world's great "floating palaces"—1890's to 1980's: Titanic, Ile de France, Queen Elizabeth, United States, Europa, more. Approx. 200 black-and-white photographs. Captions. Text. Introduction. 160pp. 8⅜ × 11¼. 24756-2 Pa. $9.95

THE GREAT LUXURY LINERS, 1927-1954: A Photographic Record, William H. Miller, Jr. Nostalgic tribute to heyday of ocean liners. 186 photos of Ile de France, Normandie, Leviathan, Queen Elizabeth, United States, many others. Interior and exterior views. Introduction. Captions. 160pp. 9 × 12. 24056-8 Pa. $9.95

A NATURAL HISTORY OF THE DUCKS, John Charles Phillips. Great landmark of ornithology offers complete detailed coverage of nearly 200 species and subspecies of ducks: gadwall, sheldrake, merganser, pintail, many more. 74 full-color plates, 102 black-and-white. Bibliography. Total of 1,920pp. 8⅜ × 11¼. 25141-1, 25142-X Cloth. Two-vol. set $100.00

THE SEAWEED HANDBOOK: An Illustrated Guide to Seaweeds from North Carolina to Canada, Thomas F. Lee. Concise reference covers 78 species. Scientific and common names, habitat, distribution, more. Finding keys for easy identification. 224pp. 5⅜ × 8½. 25215-9 Pa. $5.95

THE TEN BOOKS OF ARCHITECTURE: The 1755 Leoni Edition, Leon Battista Alberti. Rare classic helped introduce the glories of ancient architecture to the Renaissance. 68 black-and-white plates. 336pp. 8⅜ × 11¼. 25239-6 Pa. $14.95

MISS MACKENZIE, Anthony Trollope. Minor masterpieces by Victorian master unmasks many truths about life in 19th-century England. First inexpensive edition in years. 392pp. 5⅜ × 8½. 25201-9 Pa. $7.95

THE RIME OF THE ANCIENT MARINER, Gustave Doré, Samuel Taylor Coleridge. Dramatic engravings considered by many to be his greatest work. The terrifying space of the open sea, the storms and whirlpools of an unknown ocean, the ice of Antarctica, more—all rendered in a powerful, chilling manner. Full text. 38 plates. 77pp. 9¼ × 12. 22305-1 Pa. $4.95

THE EXPEDITIONS OF ZEBULON MONTGOMERY PIKE, Zebulon Montgomery Pike. Fascinating first-hand accounts (1805-6) of exploration of Mississippi River, Indian wars, capture by Spanish dragoons, much more. 1,088pp. 5⅜ × 8½. 25254-X, 25255-8 Pa. Two-vol. set $23.90

A CONCISE HISTORY OF PHOTOGRAPHY: Third Revised Edition, Helmut Gernsheim. Best one-volume history—camera obscura, photochemistry, daguerreotypes, evolution of cameras, film, more. Also artistic aspects—landscape, portraits, fine art, etc. 281 black-and-white photographs. 26 in color. 176pp. 8⅜ × 11¼. 25128-4 Pa. $12.95

THE DORÉ BIBLE ILLUSTRATIONS, Gustave Doré. 241 detailed plates from the Bible: the Creation scenes, Adam and Eve, Flood, Babylon, battle sequences, life of Jesus, etc. Each plate is accompanied by the verses from the King James version of the Bible. 241pp. 9 × 12. 23004-X Pa. $8.95

HUGGER-MUGGER IN THE LOUVRE, Elliot Paul. Second Homer Evans mystery-comedy. Theft at the Louvre involves sleuth in hilarious, madcap caper. "A knockout."—Books. 336pp. 5⅜ × 8½. 25185-3 Pa. $5.95

FLATLAND, E. A. Abbott. Intriguing and enormously popular science-fiction classic explores the complexities of trying to survive as a two-dimensional being in a three-dimensional world. Amusingly illustrated by the author. 16 illustrations. 103pp. 5⅜ × 8½. 20001-9 Pa. $2.25

THE HISTORY OF THE LEWIS AND CLARK EXPEDITION, Meriwether Lewis and William Clark, edited by Elliott Coues. Classic edition of Lewis and Clark's day-by-day journals that later became the basis for U.S. claims to Oregon and the West. Accurate and invaluable geographical, botanical, biological, meteorological and anthropological material. Total of 1,508pp. 5⅜ × 8½.
21268-8, 21269-6, 21270-X Pa. Three-vol. set $25.50

LANGUAGE, TRUTH AND LOGIC, Alfred J. Ayer. Famous, clear introduction to Vienna, Cambridge schools of Logical Positivism. Role of philosophy, elimination of metaphysics, nature of analysis, etc. 160pp. 5⅜ × 8½. (Available in U.S. and Canada only) 20010-8 Pa. $2.95

MATHEMATICS FOR THE NONMATHEMATICIAN, Morris Kline. Detailed, college-level treatment of mathematics in cultural and historical context, with numerous exercises. For liberal arts students. Preface. Recommended Reading Lists. Tables. Index. Numerous black-and-white figures. xvi + 641pp. 5⅜ × 8½.
24823-2 Pa. $11.95

28 SCIENCE FICTION STORIES, H. G. Wells. Novels, *Star Begotten* and *Men Like Gods,* plus 26 short stories: "Empire of the Ants," "A Story of the Stone Age," "The Stolen Bacillus," "In the Abyss," etc. 915pp. 5⅜ × 8½. (Available in U.S. only)
20265-8 Cloth. $10.95

HANDBOOK OF PICTORIAL SYMBOLS, Rudolph Modley. 3,250 signs and symbols, many systems in full; official or heavy commercial use. Arranged by subject. Most in Pictorial Archive series. 143pp. 8⅛ × 11. 23357-X Pa. $5.95

INCIDENTS OF TRAVEL IN YUCATAN, John L. Stephens. Classic (1843) exploration of jungles of Yucatan, looking for evidences of Maya civilization. Travel adventures, Mexican and Indian culture, etc. Total of 669pp. 5⅜ × 8½.
20926-1, 20927-X Pa., Two-vol. set $9.90

DEGAS: An Intimate Portrait, Ambroise Vollard. Charming, anecdotal memoir by famous art dealer of one of the greatest 19th-century French painters. 14 black-and-white illustrations. Introduction by Harold L. Van Doren. 96pp. 5⅜ × 8½.
25131-4 Pa. $3.95

PERSONAL NARRATIVE OF A PILGRIMAGE TO ALMANDINAH AND MECCAH, Richard Burton. Great travel classic by remarkably colorful personality. Burton, disguised as a Moroccan, visited sacred shrines of Islam, narrowly escaping death. 47 illustrations. 959pp. 5⅜ × 8½.   21217-3, 21218-1 Pa., Two-vol. set $19.90

PHRASE AND WORD ORIGINS, A. H. Holt. Entertaining, reliable, modern study of more than 1,200 colorful words, phrases, origins and histories. Much unexpected information. 254pp. 5⅜ × 8½. 20758-7 Pa. $4.95

THE RED THUMB MARK, R. Austin Freeman. In this first Dr. Thorndyke case, the great scientific detective draws fascinating conclusions from the nature of a single fingerprint. Exciting story, authentic science. 320pp. 5⅜ × 8½. (Available in U.S. only) 25210-8 Pa. $5.95

AN EGYPTIAN HIEROGLYPHIC DICTIONARY, E. A. Wallis Budge. Monumental work containing about 25,000 words or terms that occur in texts ranging from 3000 B.C. to 600 A.D. Each entry consists of a transliteration of the word, the word in hieroglyphs, and the meaning in English. 1,314pp. 6⅜ × 10.
23615-3, 23616-1 Pa., Two-vol. set $27.90

THE COMPLEAT STRATEGYST: Being a Primer on the Theory of Games of Strategy, J. D. Williams. Highly entertaining classic describes, with many illustrated examples, how to select best strategies in conflict situations. Prefaces. Appendices. xvi + 268pp. 5⅜ × 8½. 25101-2 Pa. $5.95

THE ROAD TO OZ, L. Frank Baum. Dorothy meets the Shaggy Man, little Button-Bright and the Rainbow's beautiful daughter in this delightful trip to the magical Land of Oz. 272pp. 5⅜ × 8. 25208-6 Pa. $4.95

POINT AND LINE TO PLANE, Wassily Kandinsky. Seminal exposition of role of point, line, other elements in non-objective painting. Essential to understanding 20th-century art. 127 illustrations. 192pp. 6½ × 9¼. 23808-3 Pa. $4.50

LADY ANNA, Anthony Trollope. Moving chronicle of Countess Lovel's bitter struggle to win for herself and daughter Anna their rightful rank and fortune—perhaps at cost of sanity itself. 384pp. 5⅜ × 8½. 24669-8 Pa. $6.95

EGYPTIAN MAGIC, E. A. Wallis Budge. Sums up all that is known about magic in Ancient Egypt: the role of magic in controlling the gods, powerful amulets that warded off evil spirits, scarabs of immortality, use of wax images, formulas and spells, the secret name, much more. 253pp. 5⅜ × 8½. 22681-6 Pa. $4.00

THE DANCE OF SIVA, Ananda Coomaraswamy. Preeminent authority unfolds the vast metaphysic of India: the revelation of her art, conception of the universe, social organization, etc. 27 reproductions of art masterpieces. 192pp. 5⅜ × 8½.
24817-8 Pa. $5.95

CHRISTMAS CUSTOMS AND TRADITIONS, Clement A. Miles. Origin, evolution, significance of religious, secular practices. Caroling, gifts, yule logs, much more. Full, scholarly yet fascinating; non-sectarian. 400pp. 5⅜ × 8½.
23354-5 Pa. $6.50

THE HUMAN FIGURE IN MOTION, Eadweard Muybridge. More than 4,500 stopped-action photos, in action series, showing undraped men, women, children jumping, lying down, throwing, sitting, wrestling, carrying, etc. 390pp. 7⅞ × 10⅝.
20204-6 Cloth. $21.95

THE MAN WHO WAS THURSDAY, Gilbert Keith Chesterton. Witty, fast-paced novel about a club of anarchists in turn-of-the-century London. Brilliant social, religious, philosophical speculations. 128pp. 5⅜ × 8½. 25121-7 Pa. $3.95

A CEZANNE SKETCHBOOK: Figures, Portraits, Landscapes and Still Lifes, Paul Cezanne. Great artist experiments with tonal effects, light, mass, other qualities in over 100 drawings. A revealing view of developing master painter, precursor of Cubism. 102 black-and-white illustrations. 144pp. 8¾ × 6⅜. 24790-2 Pa. $5.95

AN ENCYCLOPEDIA OF BATTLES: Accounts of Over 1,560 Battles from 1479 B.C. to the Present, David Eggenberger. Presents essential details of every major battle in recorded history, from the first battle of Megiddo in 1479 B.C. to Grenada in 1984. List of Battle Maps. New Appendix covering the years 1967–1984. Index. 99 illustrations. 544pp. 6½ × 9¼. 24913-1 Pa. $14.95

AN ETYMOLOGICAL DICTIONARY OF MODERN ENGLISH, Ernest Weekley. Richest, fullest work, by foremost British lexicographer. Detailed word histories. Inexhaustible. Total of 856pp. 6½ × 9¼.
21873-2, 21874-0 Pa., Two-vol. set $17.00

WEBSTER'S AMERICAN MILITARY BIOGRAPHIES, edited by Robert McHenry. Over 1,000 figures who shaped 3 centuries of American military history. Detailed biographies of Nathan Hale, Douglas MacArthur, Mary Hallaren, others. Chronologies of engagements, more. Introduction. Addenda. 1,033 entries in alphabetical order. xi + 548pp. 6½ × 9¼. (Available in U.S. only)
24758-9 Pa. $11.95

LIFE IN ANCIENT EGYPT, Adolf Erman. Detailed older account, with much not in more recent books: domestic life, religion, magic, medicine, commerce, and whatever else needed for complete picture. Many illustrations. 597pp. 5⅜ × 8½.
22632-8 Pa. $8.50

HISTORIC COSTUME IN PICTURES, Braun & Schneider. Over 1,450 costumed figures shown, covering a wide variety of peoples: kings, emperors, nobles, priests, servants, soldiers, scholars, townsfolk, peasants, merchants, courtiers, cavaliers, and more. 256pp. 8⅜ × 11¼. 23150-X Pa. $7.95

THE NOTEBOOKS OF LEONARDO DA VINCI, edited by J. P. Richter. Extracts from manuscripts reveal great genius; on painting, sculpture, anatomy, sciences, geography, etc. Both Italian and English. 186 ms. pages reproduced, plus 500 additional drawings, including studies for *Last Supper*, *Sforza* monument, etc. 860pp. 7⅞ × 10¾. (Available in U.S. only) 22572-0, 22573-9 Pa., Two-vol. set $25.90

THE ART NOUVEAU STYLE BOOK OF ALPHONSE MUCHA: All 72 Plates from "Documents Decoratifs" in Original Color, Alphonse Mucha. Rare copyright-free design portfolio by high priest of Art Nouveau. Jewelry, wallpaper, stained glass, furniture, figure studies, plant and animal motifs, etc. Only complete one-volume edition. 80pp. 9⅜ × 12¼.   24044-4 Pa. $8.95

ANIMALS: 1,419 COPYRIGHT-FREE ILLUSTRATIONS OF MAMMALS, BIRDS, FISH, INSECTS, ETC., edited by Jim Harter. Clear wood engravings present, in extremely lifelike poses, over 1,000 species of animals. One of the most extensive pictorial sourcebooks of its kind. Captions. Index. 284pp. 9 × 12.
                                                                 23766-4 Pa. $9.95

OBELISTS FLY HIGH, C. Daly King. Masterpiece of American detective fiction, long out of print, involves murder on a 1935 transcontinental flight—"a very thrilling story"—NY Times. Unabridged and unaltered republication of the edition published by William Collins Sons & Co. Ltd., London, 1935. 288pp. 5⅜ × 8½. (Available in U.S. only)   25036-9 Pa. $4.95

VICTORIAN AND EDWARDIAN FASHION: A Photographic Survey, Alison Gernsheim. First fashion history completely illustrated by contemporary photographs. Full text plus 235 photos, 1840-1914, in which many celebrities appear. 240pp. 6½ × 9¼.   24205-6 Pa. $6.00

THE ART OF THE FRENCH ILLUSTRATED BOOK, 1700-1914, Gordon N. Ray. Over 630 superb book illustrations by Fragonard, Delacroix, Daumier, Doré, Grandville, Manet, Mucha, Steinlen, Toulouse-Lautrec and many others. Preface. Introduction. 633 halftones. Indices of artists, authors & titles, binders and provenances. Appendices. Bibliography. 608pp. 8⅜ × 11¼.   25086-5 Pa. $24.95

THE WONDERFUL WIZARD OF OZ, L. Frank Baum. Facsimile in full color of America's finest children's classic. 143 illustrations by W. W. Denslow. 267pp. 5⅜ × 8½.   20691-2 Pa. $5.95

FRONTIERS OF MODERN PHYSICS: New Perspectives on Cosmology, Relativity, Black Holes and Extraterrestrial Intelligence, Tony Rothman, et al. For the intelligent layman. Subjects include: cosmological models of the universe; black holes; the neutrino; the search for extraterrestrial intelligence. Introduction. 46 black-and-white illustrations. 192pp. 5⅜ × 8½.   24587-X Pa. $6.95

THE FRIENDLY STARS, Martha Evans Martin & Donald Howard Menzel. Classic text marshalls the stars together in an engaging, non-technical survey, presenting them as sources of beauty in night sky. 23 illustrations. Foreword. 2 star charts. Index. 147pp. 5⅜ × 8½.   21099-5 Pa. $3.50

FADS AND FALLACIES IN THE NAME OF SCIENCE, Martin Gardner. Fair, witty appraisal of cranks, quacks, and quackeries of science and pseudoscience: hollow earth, Velikovsky, orgone energy, Dianetics, flying saucers, Bridey Murphy, food and medical fads, etc. Revised, expanded In the Name of Science. "A very able and even-tempered presentation."—The New Yorker. 363pp. 5⅜ × 8.
                                                                 20394-8 Pa. $6.50

ANCIENT EGYPT: ITS CULTURE AND HISTORY, J. E Manchip White. From pre-dynastics through Ptolemies: society, history, political structure, religion, daily life, literature, cultural heritage. 48 plates. 217pp. 5⅜ × 8½.   22548-8 Pa. $4.95

SIR HARRY HOTSPUR OF HUMBLETHWAITE, Anthony Trollope. Incisive, unconventional psychological study of a conflict between a wealthy baronet, his idealistic daughter, and their scapegrace cousin. The 1870 novel in its first inexpensive edition in years. 250pp. 5⅜ × 8½. 24953-0 Pa. $5.95

LASERS AND HOLOGRAPHY, Winston E. Kock. Sound introduction to burgeoning field, expanded (1981) for second edition. Wave patterns, coherence, lasers, diffraction, zone plates, properties of holograms, recent advances. 84 illustrations. 160pp. 5⅜ × 8¼. (Except in United Kingdom) 24041-X Pa. $3.50

INTRODUCTION TO ARTIFICIAL INTELLIGENCE: SECOND, ENLARGED EDITION, Philip C. Jackson, Jr. Comprehensive survey of artificial intelligence—the study of how machines (computers) can be made to act intelligently. Includes introductory and advanced material. Extensive notes updating the main text. 132 black-and-white illustrations. 512pp. 5⅜ × 8½. 24864-X Pa. $8.95

HISTORY OF INDIAN AND INDONESIAN ART, Ananda K. Coomaraswamy. Over 400 illustrations illuminate classic study of Indian art from earliest Harappa finds to early 20th century. Provides philosophical, religious and social insights. 304pp. 6⅝ × 9⅝. 25005-9 Pa. $8.95

THE GOLEM, Gustav Meyrink. Most famous supernatural novel in modern European literature, set in Ghetto of Old Prague around 1890. Compelling story of mystical experiences, strange transformations, profound terror. 13 black-and-white illustrations. 224pp. 5⅜ × 8½. (Available in U.S. only) 25025-3 Pa. $5.95

ARMADALE, Wilkie Collins. Third great mystery novel by the author of *The Woman in White* and *The Moonstone*. Original magazine version with 40 illustrations. 597pp. 5⅜ × 8½. 23429-0 Pa. $9.95

PICTORIAL ENCYCLOPEDIA OF HISTORIC ARCHITECTURAL PLANS, DETAILS AND ELEMENTS: With 1,880 Line Drawings of Arches, Domes, Doorways, Facades, Gables, Windows, etc., John Theodore Haneman. Sourcebook of inspiration for architects, designers, others. Bibliography. Captions. 141pp. 9 × 12. 24605-1 Pa. $6.95

BENCHLEY LOST AND FOUND, Robert Benchley. Finest humor from early 30's, about pet peeves, child psychologists, post office and others. Mostly unavailable elsewhere. 73 illustrations by Peter Arno and others. 183pp. 5⅜ × 8½. 22410-4 Pa. $3.95

ERTÉ GRAPHICS, Erté. Collection of striking color graphics: *Seasons, Alphabet, Numerals, Aces* and *Precious Stones*. 50 plates, including 4 on covers. 48pp. 9⅜ × 12¼. 23580-7 Pa. $6.95

THE JOURNAL OF HENRY D. THOREAU, edited by Bradford Torrey, F. H. Allen. Complete reprinting of 14 volumes, 1837–61, over two million words; the sourcebooks for *Walden*, etc. Definitive. All original sketches, plus 75 photographs. 1,804pp. 8½ × 12¼. 20312-3, 20313-1 Cloth., Two-vol. set $80.00

CASTLES: THEIR CONSTRUCTION AND HISTORY, Sidney Toy. Traces castle development from ancient roots. Nearly 200 photographs and drawings illustrate moats, keeps, baileys, many other features. Caernarvon, Dover Castles, Hadrian's Wall, Tower of London, dozens more. 256pp. 5⅜ × 8¼. 24898-4 Pa. $5.95

AMERICAN CLIPPER SHIPS: 1833–1858, Octavius T. Howe & Frederick C. Matthews. Fully-illustrated, encyclopedic review of 352 clipper ships from the period of America's greatest maritime supremacy. Introduction. 109 halftones. 5 black-and-white line illustrations. Index. Total of 928pp. 5⅜ × 8½.
25115-2, 25116-0 Pa., Two-vol. set $17.90

TOWARDS A NEW ARCHITECTURE, Le Corbusier. Pioneering manifesto by great architect, near legendary founder of "International School." Technical and aesthetic theories, views on industry, economics, relation of form to function, "mass-production spirit," much more. Profusely illustrated. Unabridged translation of 13th French edition. Introduction by Frederick Etchells. 320pp. 6⅛ × 9¼. (Available in U.S. only) 25023-7 Pa. $8.95

THE BOOK OF KELLS, edited by Blanche Cirker. Inexpensive collection of 32 full-color, full-page plates from the greatest illuminated manuscript of the Middle Ages, painstakingly reproduced from rare facsimile edition. Publisher's Note. Captions. 32pp. 9⅜ × 12¼. 24345-1 Pa. $4.95

BEST SCIENCE FICTION STORIES OF H. G. WELLS, H. G. Wells. Full novel *The Invisible Man,* plus 17 short stories: "The Crystal Egg," "Aepyornis Island," "The Strange Orchid," etc. 303pp. 5⅜ × 8½. (Available in U.S. only)
21531-8 Pa. $4.95

AMERICAN SAILING SHIPS: Their Plans and History, Charles G. Davis. Photos, construction details of schooners, frigates, clippers, other sailcraft of 18th to early 20th centuries—plus entertaining discourse on design, rigging, nautical lore, much more. 137 black-and-white illustrations. 240pp. 6⅛ × 9¼.
24658-2 Pa. $5.95

ENTERTAINING MATHEMATICAL PUZZLES, Martin Gardner. Selection of author's favorite conundrums involving arithmetic, money, speed, etc., with lively commentary. Complete solutions. 112pp. 5⅜ × 8½. 25211-6 Pa. $2.95

THE WILL TO BELIEVE, HUMAN IMMORTALITY, William James. Two books bound together. Effect of irrational on logical, and arguments for human immortality. 402pp. 5⅜ × 8½. 20291-7 Pa. $7.50

THE HAUNTED MONASTERY and **THE CHINESE MAZE MURDERS,** Robert Van Gulik. 2 full novels by Van Gulik continue adventures of Judge Dee and his companions. An evil Taoist monastery, seemingly supernatural events; overgrown topiary maze that hides strange crimes. Set in 7th-century China. 27 illustrations. 328pp. 5⅜ × 8½. 23502-5 Pa. $5.95

CELEBRATED CASES OF JUDGE DEE (DEE GOONG AN), translated by Robert Van Gulik. Authentic 18th-century Chinese detective novel; Dee and associates solve three interlocked cases. Led to Van Gulik's own stories with same characters. Extensive introduction. 9 illustrations. 237pp. 5⅜ × 8½.
23337-5 Pa. $4.95

Prices subject to change without notice.

Available at your book dealer or write for free catalog to Dept. GI, Dover Publications, Inc., 31 East 2nd St., Mineola, N.Y. 11501. Dover publishes more than 175 books each year on science, elementary and advanced mathematics, biology, music, art, literary history, social sciences and other areas.